A PLUME BOOK

WORRY PROOF

DR. CARA NATTERSON is a graduate of Harvard and the Johns Hopkins School of Medicine. She lives in Los Angeles with her husband and two children.

"A word of warning. If you are a hypochondriac-by-proxy who, like me, enjoys freaking out about every imagined danger to her children, do NOT buy this book. Dr. Natterson's sage and reasoned advice will dispel the perils that plague your daydreams, while making clear what you should, in fact, be concerned about. It will, quite simply, calm you down and thus ruin your neurotic day. If, on the other hand, you would actually appreciate sensible and knowledgeable advice, buy this book immediately."
—Ayelet Waldman, author of *Bad Mother: A Chronicle of Maternal Crimes, Minor Calamities, and Occasional Moments of Grace*

. Cara Natterson cuts through
tion that confuses and worries
s and what doesn't."
:neral Pediatrics and William
n's Hospital Boston/Harvard
*You Never Wanted Your Kids to
: (But Were Afraid They'd Ask)*

"Dr. Natterson aims to put to rest 'unfounded and overhyped fears' that seem epidemic among today's parents who are bombarded with tales of the 'dangers' lurking in every corner of their home. This book is worth the chapter on vaccines alone!"
—Lara Zibners, MD, author of *If Your Kid Eats This Book, Everything Will Still Be Okay*

"Dr. Natterson sets the records straight for parents regarding what is safe for their children to put in and on their bodies. . . . Science-based and presented in an easy to understand manner. Her focus is to educat
about what foods and products might e
highly recommend this book for my pat
—Gary Rachelefsky, MD, auth and Pro

Worry Proof

A Pediatrician (*and Mom*)
Explains Which Foods,
Medicines, and Chemicals
to Avoid to Have Safe
and Healthy Children

Cara Natterson, MD

Previously published as *Dangerous or Safe?*

A PLUME BOOK

PLUME
Published by the Penguin Group
Penguin Group (USA) Inc., 375 Hudson Street, New York, New York 10014, U.S.A. •
Penguin Group (Canada), 90 Eglinton Avenue East, Suite 700, Toronto, Ontario, Canada
M4P 2Y3 (a division of Pearson Penguin Canada Inc.) • Penguin Books Ltd., 80 Strand,
London WC2R 0RL, England • Penguin Ireland, 25 St. Stephen's Green, Dublin 2, Ireland
(a division of Penguin Books Ltd.) • Penguin Group (Australia), 250 Camberwell Road,
Camberwell, Victoria 3124, Australia (a division of Pearson Australia Group Pty. Ltd.) •
Penguin Books India Pvt. Ltd., 11 Community Centre, Panchsheel Park, New Delhi – 110
017, India • Penguin Group (NZ), 67 Apollo Drive, Rosedale, North Shore 0632, New
Zealand (a division of Pearson New Zealand Ltd.) • Penguin Books (South Africa) (Pty.)
Ltd., 24 Sturdee Avenue, Rosebank, Johannesburg 2196, South Africa

Penguin Books Ltd., Registered Offices: 80 Strand, London WC2R 0RL, England

Published by Plume, a member of Penguin Group (USA) Inc. Previously published in a
Hudson Street Press edition as *Dangerous or Safe?*

First Plume Printing, November 2010

10 9 8 7 6 5 4 3 2 1

 REGISTERED TRADEMARK—MARCA REGISTRADA

The Library of Congress has catalogued the Hudson Street Press edition as follows:

Natterson, Cara Familian, 1970–
Dangerous or safe? : which foods, medicines, and chemicals really put your kids at risk / Cara
Natterson.
   p. cm.
Includes bibliographical references and index.
ISBN 978-1-59463-062-0 (hc.)
ISBN 978-0-452-29659-6 (pbk.)
1. Pediatric toxicology—Popular works.
I. Title.
RA1225.N384 2009
618.92'98—dc22   2009019782

Printed in the United States of America
Set in Janson Text with Helvetica • Original hardcover design by Daniel Lagin

PUBLISHER'S NOTE
Every effort has been made to ensure that the information contained in this book is complete
and accurate. However, neither the publisher nor the author is engaged in rendering
professional advice or services to the individual reader. The ideas, procedures, and
suggestions contained in this book are not intended as a substitute for consulting with your
physician. All matters regarding your health require medical supervision. Neither the
author nor the publisher shall be liable or responsible for any loss or damage allegedly
arising from any information or suggestion in this book.

For Paul, Talia, and Ry

Acknowledgments

THIS book took shape during years of clinical practice, in exam rooms and at conferences. But it became a real entity along the sidelines of the soccer field where my kids play after school. Thanks to all the parents who asked questions and particularly to the moms from Circle of Children—Suzy Feldman, Maya Pinkner, Laura Fox, Missy Polson, Amy Hanning, Amy Listen—who sat in the grass week after week sharing the latest hype. Also to Michele Gathrid, Carrie Casden, and Aleksandra Crapanzano, who read through drafts along the way, and to Lindsey Kozberg who got me away from writing when I needed a break.

To my guides and pillars: Richard Abate, Trena Keating, and Nancy Josephson. To Bryan Wolf and Jamie Afifi at Ziffren. And to Luke Dempsey and Meghan Stevenson at Hudson Street Press. Thanks to my mentors and friends at Tenth Street Pediatrics, especially Bill Gurfield and Jim Varga (who peppered me with science throughout the writing of this book) and Lisa Stern (the sister I never had). I am also grateful to the women of Telepictures—particularly Sheila Bouttier, Lisa Hackner, and Hilary Estey McLoughlin—who provided me a new home outside of clinical practice.

To my own mom, who is an extraordinary parent without breaking a sweat. To Talia, Ry, and Rosa, who fill my home with happiness. And to my husband Paul, the love of my life and the most risk-averse person I know.

Contents

Introduction

What Does "Dangerous" Really Mean?

OVER and over we ask ourselves a simple question: is it dangerous or safe? We wonder, should I take this medicine? Eat this food? Buy this product? All day, every day, we make mental calculations that boil down to this black-and-white question: dangerous or safe? When we ask this question about ourselves, it is usually easy to answer. But when it comes to our children, nothing seems clear, and the gray zone feels enormous. For ourselves, we may be comfortable accepting uncertainty when the answer isn't obvious. But for our kids we are not: for them we need concrete, clear answers without risk or ambiguity. When it comes to our children, we worry more.

I suppose it's human nature—we simply don't have enough brain space to worry about everything for everyone all the time. So we pour our angst into our highest priority, our children. Should they be drinking out of plastic bottles, using cell phones, eating processed foods, taking antibiotics, receiving vaccines? Some of these questions have real answers; others just hype. Regardless, we parents have hit a point where we torture ourselves over every detail of our children's lives—all while sipping from our own plastic bottles, talking on cell phones, snacking on processed foods, and swallowing medications.

I am the mother of two young children. This means that I belong to the parenting generation that has been accused of being overbearing, worrying about every little thing, and trying to control every aspect of our children's lives. Generally the accusers are our own parents. "You survived childhood," they say, in a slightly mocking tone, "and we never worried as much about every little thing as you do."

This is true. But our parents lived in a very different world. When our parents had young kids, information was largely limited to the newspaper and the evening news. As a result, the news focused on the most important issues of the day. Today news is a constant barrage that includes twenty-four-hour cable networks, live Web streaming, and anything that might fill a few minutes of screen time or a few inches of crawl space along the bottom of your TV. This lends itself to Breaking Alerts! about pediatric-health horror stories: "Child stops speaking after receiving a vaccine!" "Flesh-eating bacteria spreading through school community!" And then there are the headlines about product recalls: "Don't Give Your Child a Toy Train Because the Paint Is Leaded!" "Don't Let Your Child Sleep in Flame-Retardant Pajamas Because They Are Toxic!" With these arriving on a weekly, sometimes daily, basis, how can we possibly be expected to ignore them?

Beyond being a mom, I am also a pediatrician. A big part of the job is fielding phone calls from worried parents. With each breaking news story, parents want to know what to do for their child. Some take the time to do research on their own, but most people have learned that if you Google long enough, you'll find two sides to every story. This leaves parents even more confused than they were at the outset. So they call the doctor, looking for the simple yes or no answer. When do I need to worry? That's all parents really want to know.

It is ironic that despite our need for simple, straightforward answers we crave more and more information. So much news is coming at us all the time, but most people have no sense of what to do with it, how to prioritize it, and when to worry about it. This is certainly not to say that information should be kept from the public. But as a result of the onslaught, we begin to fear that danger lurks at every turn. With so much to consider, it is easy to lose sight of both the true and relative risks.

Relative risk simply means the risk of an event occurring in connection to an exposure. If one group of people is exposed to something and another is not, the relative risk is the probability that the exposed group will have a specific outcome. In medicine, that outcome may mean developing a disease or even dying.

Risk assessment is a calculation we make many times every day without even knowing it. What is the chance that I will be hit by a car if I jaywalk? What is the chance that I will get a sunburn (or one day even skin cancer) if I don't put on that sunscreen? What is the chance that I will be late to work if I roll over for five or ten more minutes of sleep? Risk assessment can be applied to every decision in our daily life, down to the most mundane.

Relative risk can also be used in a broad sense, forcing us to step away from the trees and look at the whole forest. There are things in our world that are relatively more dangerous than others. For instance, playing with a loaded handgun is a heck of a lot more dangerous than taking a break to get a drink of water out of a plastic bottle. We all know this—no one would disagree. But millions of Americans keep guns in their homes, loaded and accessible to their children.[1] This may seem like a ridiculous example, but during the past few years the debate over the safety of plastics has been a continuously covered news item while guns in the home rarely make headlines. Ultimately, what we read about in the paper or online, hear about on TV, and talk about with friends tends to be in the forefront of our minds, often magnifying the actual risk. These days, because we are increasingly focused on specific issues, we may overlook things that are relatively more dangerous.

Whether we agonize over the foods we eat or the chemicals in our environment, it is easy to lose sight of the actual number of people affected in a negative way. When we blow potential hazards out of proportion, we think intently about tiny decisions and start to see much of our world through a narrow lens. Many parents tell me they don't like approaching the world this way but they just can't help it.

The inspiration for this book came from my desire to unburden parents while also educating them about what is truly dangerous for their kids (and themselves). There is good scientific data available out there; unfortunately it is often difficult to decipher unless you are trained to read medical articles. That's why you are reading this book: so that you can understand where the hype ends and where the truth begins, so you can learn to identify what might really endanger your kids and then be able to avoid those hazards like plagues.

The fact is that we *do* have to worry about every little thing more than our parents did, because life has changed. Since we were children, many new chemicals have been invented. In addition, many materials once used sparingly have now become ubiquitous. Just look at plastics. Phthalates and bisphenol A—chemicals used to increase the functionality of plastics—aren't new, but they are now used in thousands of household items. When a chemical is utilized in every corner of our lives, it is reasonable to ask whether it is dangerous or safe.

A generation ago, a medicine or chemical needed to have catastrophic consequences (like birth defects, cancers, or deaths) to qualify as "dangerous." Now far subtler outcomes are analyzed. Does something cause a fall in test scores, depression, premature breast development, or acne? We remain concerned about disastrous results, but we also want to avoid even the most minor repercussions. This is why it is fair to ask these questions—just because something doesn't have catastrophic consequences doesn't necessarily mean it's safe.

Worry Proof examines the foods we eat, liquids we drink, chemicals in our environment, and medicines we take. I do not need to write a chapter about why my children will never be allowed to ride on motorcycles, because the answer is obvious. This book concentrates on the cloudy waters of the more subtle questions. It translates the data and provides clear answers. It is not meant to perpetuate drama but rather quite the opposite—to put to rest unfounded and overhyped fears.

There is an army of devoted physicians, scientists, academics, consultants, and journalists studying how exposures to various chemicals may or may not affect our future health. There are people who look at issues on the microscopic level, literally studying one cell or a single chemical reaction, and there are others who analyze how a food or piece of technology or a lab-manufactured additive impacts an entire population. *Worry Proof* takes the current data and distills it down to its core facts so that you get the bottom line: what is safe and what is not.

There *are* answers to questions about whether many of the things we use regularly are safe. The chapters that follow provide you with concrete evidence and advice. To do so, each chapter is organized into four sections:

What Is the Question?
What Is the Data?
What Is the Bottom Line?
What's in My Home?

The question section takes a broad issue (like plastic bottles) and defines it more specifically (is bisphenol A really dangerous?). The data section summarizes the history of the question and then provides a translation of scientific literature into layman's terms. I use published data from reputable journals and peer-reviewed articles. I also provide government data from the NIH, CDC, FDA, and other agencies because these materials form the basis for many of the guidelines and policies currently in effect. Since the abbreviations and terminology can be confusing, I've put the acronyms and medical terms in bold print the first time they appear within a chapter; any term in bold is defined in the glossary at the back. In the bottom-line section, I offer my own opinion, as a mom and as a pediatrician, about whether something is dangerous or safe. And finally, I answer the question that all my patients and friends ask: so, what do you do with *your* kids?

Worry Proof will give you many answers, but I feel compelled to point out that because there are new scientific discoveries every day it is possible that something we think is safe today may turn out to be harmful tomorrow. Even in the year between the hardcover and paperback publications of this book, some of the data has been changed. This updated version of the book reflects new pieces of information. They said, if you understand the data and tune out the hype, the chance that you will expose your children (or yourself) to something truly dangerous is actually quite low. Many people dedicate their lives to researching these questions and answering them as truthfully as possible. I hope *Worry Proof* will help you understand what experts have learned and allow you to make smart decisions about your family's health and safety.

Part I

Foods

Chapter 1

Allergens

WHAT IS THE QUESTION?

A generation ago, food allergies were the exception rather than the rule. I remember the headline when I was in college that a sophomore at a nearby school had died from a peanut allergy. She had been out to dinner and had taken a bite of chili that, unbeknownst to her, had some peanuts in it. There was another front-page story, when I was in pediatrics training, about a girl who died kissing her boyfriend. He had eaten peanuts that day and inadvertently poisoned her. These events were shocking. Who knew anyone who had a peanut allergy, let alone had died from it?

Today the stories seem much less surprising. There are kids everywhere with food allergies. The types of allergies include everything from nuts to wheat, cow's milk to soy, citrus to strawberry, shellfish to chicken. Almost every school has at least one seriously food-allergic child. This necessitates making rules about what other children bring for lunch or setting up allergy-free tables to keep the allergic kids safe.

This isn't an imaginary trend—it's an explosion of a once rare problem. Food allergy is now seen in 4 percent of all kids in the United States (Branum 2008). The rates of hospitalization for food allergy and anaphylaxis (the dreaded life-threatening complication of an allergic reaction) are increasing. And there is no sign of slowing. Population studies suggest that the prevalence of food allergies will continue to grow rapidly.

In fact, even since the initial printing of this book, the reported numbers have increased again. The American Academy of Allergy

Asthma and Immunology (AAAAI) states that now 6 percent of children under the age of three have food allergy, and 3–4 percent of adults do. Cases of anaphylaxis from food have increased from 21,000 per year in 1999 to 51,000 per year in 2008. Food allergy–related hospital admissions have increased from 2,600 per year in 1998–2000 to 9,500 per year in 2004–2006.*
    Why are we increasingly allergic to our food supply? And what can we do to help our children avoid this fate?

WHAT IS THE DATA?

Food allergies affect almost three million U.S. children—that's nearly 4 percent of all people under age eighteen. And the number keeps on growing. Some reports estimate that in the United States the number of peanut allergies alone doubled between 1997 and 2002 (Sicherer 2003).
    Just about any food can trigger a reaction. However, eight foods are responsible for most of these allergies: cow's milk, soybeans, eggs, wheat, fish, shellfish, tree nuts,[1] and, of course, peanuts.
    Severe allergic reactions are those that can cause a life-threatening illness called anaphylaxis. When anaphylactic shock occurs, there is a massive release of inflammatory cells called mast cells. The mast cells unload their native chemicals (with names like histamines, leukotrienes, and prostaglandins), which in turn trigger inflammation all over the body. These chemicals cause blood vessels to dilate and blood pressure to drop; generate swelling in the lining of the airways, which creates difficulty breathing; can cause swelling in the gastrointestinal tract, precipitating cramps and vomiting; and inflame the lips, eyes, and skin, creating a hivelike rash.[2]
    Anaphylactic shock is the most worrisome consequence of food allergy and can be fatal. The anaphylaxis usually occurs within one to two hours of ingesting the offending food, but it can happen within seconds. There are between one hundred and two hundred deaths every year in this country as a direct result of food allergies. Adolescents and young adults are at the highest risk.
    In the United States, peanuts and tree nuts cause the majority of

---

* www.aaaai.org/media/statistics/allergy-statistics.asp

severe allergic reactions with anaphylactic shock. Fish and shellfish can also be culprits, but this happens less frequently.[3] Many children are aware of their food allergies and know to avoid certain things. Still, accidental exposure is common. One study showed that 58 percent of kids diagnosed with peanut allergy were accidentally exposed within five years of the diagnosis (Shah 2008). A different study found that one-third of peanut- or tree nut–related fatalities were caused when a person ate a dessert prepared outside the home (Bock 2007).[4]

So why are there more allergies today than there were twenty or thirty years ago? There is more than one answer to that question. One explanation has to do with overlapping diagnoses. Doctors have only recently fully appreciated that a person with one type of allergy is likely to have other allergic illnesses. A dry, allergic rash on the skin is called eczema. Allergy in the lungs is called asthma, or in its milder form "reactive airways disease." In medicine the triad of allergy, eczema, and asthma is called atopy. Because the three diagnoses within atopy travel together, doctors have learned to look for all of them. As a result, it is not uncommon for an asthmatic child to undergo food-allergy testing even if the child has never exhibited obvious signs of sensitivity to specific foods. While this approach identifies and prevents some problems, it also inflates the number of diagnoses, because children are being identified with an issue before they ever have a symptom.

The problems also filter down through family members. If a parent has atopy, there is a 25 percent chance that the child will develop food allergy before the age of seven. If a child has a first-degree family member with peanut allergy, that child's risk of peanut allergy increases sevenfold (Kumar 2008).

Another reason why there are more food allergies today is semantic: we are labeling more symptoms as "food allergies" than we did in the past. Today children with food allergies are typically diagnosed between one and three years of age. But by the time these kids get to middle childhood—second or third grade—half of them are no longer considered food allergic. In the past, many of these children who had mild symptoms were simply not diagnosed and, as they got older, outgrew the problem. Kids are still outgrowing the problem today, but more of them are being diagnosed with a food allergy when they are young.

There are also many kids who are mislabeled as food allergic. This group may have food intolerance but not true allergy. Food intolerance means that the food elicits a symptom—like nausea or vomiting or even rash—but the reaction is not caused by the immune system. It is technically not an allergy. Examples of food intolerance include lactose intolerance, gallbladder disease (in which fat is not tolerated), food poisoning (a temporary but intense food intolerance), headache following the ingestion of tyramine (found in aged cheeses, avocado, and soy products), jitters or palpitations with caffeine, and so on.

The increase in food-allergy diagnosis is not just a function of terminology. Yes, there are kids called allergic today who would have been overlooked in the past. But when you factor them out of the equation, there *really are* more kids who are food allergic today than there were in past generations. Peanut allergies have made the most dramatic leap: A study published in 2010 showed that 1.4 percent of kids now have allergy to peanuts, more than triple the number from ten years prior.

So why do so many kids seem to have allergies these days? That question brings us to the final reason: something is triggering them. And lots of people are trying to figure out what that trigger is.

One theory, called the hygiene hypothesis, blames the immune system for the increase in allergies. According to this theory, the maturation of a newborn's immune system relies heavily upon its interaction with normal infections (like bacteria, viruses, and parasites). The argument is that if a parent keeps things too clean or shields a baby from "normal" germs, the infant's immune system doesn't come into contact with infectious agents early enough. That baby's immune system will develop an inappropriate sense of what are foreign invaders—so inappropriate, in fact, that certain foods may be seen as foreign, stimulating the immune system and prompting an allergic response. The body's response to a food allergen is not much different from its reaction to a common cold virus or a dirty shard of glass in the foot. In each of these cases, the immune system thinks that it has to get to work to protect the body from a foreign invader, and that involves massive inflammation.

The hygiene hypothesis has many critics, most of whom say that while allergy may be a symptom of an inappropriate immune system response it has nothing to do with infections and germs. Many of these critics contend that food allergy is a result of exposure to a culprit food, a phenomenon called sensitization. There is a schism within the group of people who believe in sensitization. One camp argues that a child is more likely to become allergic to a food if the food is offered too early in life; the other says the exact opposite: that offering the food allows the immune system to recognize it and to wait too long can create the allergy.

The **AAP**'s guidelines about the introduction of solid foods for infants both recognize and sidestep the debate over food allergy. If allergy is related to the introduction of food, what is the best age to start and what should you give first? Experts suggest that solids be started no earlier than four months of age and no later than six months, and recommend that fruits and vegetables be introduced before proteins.[5] Grains in the form of infant rice cereal are also acceptable starter foods. But it remains completely unclear whether the AAP believes that the risk of food allergy is increased (or potentially even decreased) by exposure to certain foods in a certain order over the first few months that a child is eating solids.[6]

Some people believe strongly that a mother's diet during pregnancy or breast-feeding affects the allergy status of a child. To this point, the AAP has said there is no proof that avoiding specific foods during pregnancy or lactation prevents atopy. In other words, it shouldn't matter what the mother is eating. But many doctors and parents of allergic children disagree.

One thing a mother can do to help avoid allergy in her baby is to breast-feed exclusively. Breast-fed babies are less likely to have eczema and wheezing than their formula-fed peers. But the allergy doesn't go away—it is only delayed and few agree upon why this delay occurs. Ultimately, if you are going to be allergic, you are going to be allergic—even exclusively breast-fed babies who have no eczema or wheezing may get these as they grow older. There are also plenty of breast-fed babies out there who don't experience the delay and have quite a bit of eczema or wheezing. Some of my most allergic patients were exclusively breast-fed. So unfortunately, this strategy doesn't work for everyone.

Where does this leave us in the quest to figure out why food-allergy rates are climbing? We should look for what has changed over the past few years in our food supply that might account for this shift. Here's where we turn to the science-fiction-like topic of **genetically modified foods**. Much of this information also appears in the chapter about soy, but I think it bears repeating: first, because genetic modification affects almost every one of us; and, second, because it's tough to follow the mechanics of it on the first read-through.

Over the past two decades, agricultural scientists have genetically modified the DNA of a variety of agricultural products, such as soy, corn, cottonseed, canola, tomatoes, potatoes, squash, carrots, sugar beets, papaya, wheat, rice, and milk and dairy products. The genetically modified products have been available for purchase in the United States since 1996.

The program started for laudable reasons: genetically modified crops would be able to increase the earth's bounty, preventing malnutrition or starvation in an expanding world population; create plants that can survive extreme environmental conditions like drought or freezing cold; arm plants with the ability to ward off diseases; and encode plant DNA with a natural pesticide that makes pests less likely to nibble on them and allows crops to grow without being sprayed with industrial chemicals. There are even some genetic modifiers that can be used to increase the nutritional value of a plant. In countries where the staple agricultural product is rice, for instance, genetic modification can be used to enhance the nutritional value of rice by increasing its vitamin and mineral content.

But there are two sides to every coin. Genetic modification can also be used for less-admirable purposes. For instance, some genetically modified plants survive the spraying of pesticides or herbicides. In the past, pesticides killed not only pests but also many of the plants sheltering them. Crops might produce a lower yield because of pests, but might also produce a lower yield because of pesticide sprays. With the advent of genetic modification, plants can be sprayed, the pests will die, but the plants will continue to thrive (covered, nonetheless, in pesticide). This ultimately encourages the use of pesticides.

Genetic modification has also had some unintended and particularly negative consequences. It looks like genetic modification may directly harm other organisms. For example, when monarch butterfly caterpillars are exposed to pollen from genetically modified corn, the species has much higher mortality rates. There are also growing concerns about weeds after data showed that cross-pollination between crops and weeds may create "super weeds" that are able to tolerate the sprays once used to control them.

The process of genetic modification is complex. It doesn't involve adding just one gene into the mix, but rather three or four genes (and sometimes more). Here's how it works. In order for a specific gene, say a gene for pesticide resistance, to be artificially inserted into plant DNA, three things need to happen. First, scientists have to design the new pesticide-resistant gene. This is pretty easy.

Next, the pesticide-resistant gene needs help getting into the genome—it cannot just jump into the DNA. So it is coupled with another gene, called a "promoter" gene, to help with integration. Promoter genes are particularly good at sliding into other DNA. Here's a little catch: promoter genes tend to be taken from different foods. (More on this later.)

Finally, in the third step, yet another gene is needed as a "marker." This helps scientists identify which plants were successfully modified. One common marker gene is an antibiotic-resistant gene. So in our example, when scientists want to determine whether they have successfully integrated the new pesticide-resistant gene into its recipient DNA (say, for instance, a soybean), they just sprinkle the soybean with antibiotics. If the soy survives, it has resisted the antibiotics. The antibiotic resistance acts as a marker of successful genetic modification.[7]

What does any of this have to do with food allergies? Well, genes are responsible for making proteins: when a part of the DNA is "turned on," it encodes a variety of proteins, which in turn tell the cells what to do. Genetically modified foods have a combination of standard genes and new genes in their DNA, and these new genes have the potential to produce new proteins. If the part of the DNA that contains the new genes is "turned on," the proteins produced by these novel genes could look to the human immune system like foreign invaders. When this happens, they can trigger allergic reactions. So if you eat genetically modified soy and the new genes used to modify the soy produce a specific protein that your body thinks is a foreign invader or an allergen, your immune system will rev up and you may have an allergic reaction.

This is just one theory of how genetic modification causes allergies, and nobody knows whether it is true. Another theory is that one of the genes inserted into that modified DNA (in our example, soy) belongs to a family of allergenic foods. Remember, scientists don't just invent new genes but take genes from one plant and insert them into another. In some cases, genes from allergenic foods like peanuts are used as genetic modifiers. So if a peanut-derived gene is inserted into soy *and* it can produce a protein *and* the protein it produces looks similar to the protein that causes a peanut allergy in a particular person, then that soy could trigger a serious allergic reaction when it is consumed.

This second theory sounds pretty unlikely. Only a tiny part of the foreign (in our scenario, peanut) DNA is inserted into the recipient food (our soybean). It's unlikely that the small piece of peanut DNA would be able to produce a whole protein or, if it is, the likelihood that the protein looks just like peanut allergen is small. But it is still possible. In fact, it actually happened. This was in the mid-1990s when a soybean that never made it to market was modified with a gene from the Brazil nut. Lab tests done on people allergic to Brazil nuts showed allergic reactions to the soybeans. It wasn't clear what the particular mechanism for allergy was, but the allergic response was dramatic. Had this product made it to market, it is possible that a person allergic to Brazil nuts could have developed anaphylaxis from eating a soybean (Nordlee 1996).

There is a third theory about how genetically modified foods trigger allergies, one that has very little to do with DNA alteration. It points a finger, instead, at bacteria or viruses used in the process of genetic modification. These infections can be used as promoters. In many instances they have been effective tools for the insertion of new genes (like the pesticide-resistant gene in our example) into a host DNA. Once a bacteria or virus successfully integrates itself into DNA, it carries on doing its regular thing and producing its own native proteins. Those proteins are foreign to your body. Remember, the immune system is designed to identify and reject foreign invaders. So if a bacteria or virus in a genetically modified food produces proteins, from your immune system's point of view the food (with its piggybacked virus or bacteria) is a foreign invader and its presence triggers a reaction.

Agricultural scientists don't take allergy concerns lightly. When bacteria is used in the process of modification, for instance, researchers look at the proteins the bacteria may produce and cross-reference the results with a database of proteins known to cause allergies. Based on this, **WHO** has established criteria for what modified crops should or should not be allowed on the commercial market. Even though we don't know precisely how genetic modification provokes allergic reactions, there is some good evidence that this new technology contributes to higher food-allergy rates.

This is the point in the story when most people stop and ask: How big a problem is this? "Genetically modified" sounds so high-tech—it cannot possibly affect a large percentage of the foods in my local grocery store. Wrong. As of 2008, 92 percent of the soy, 80 percent of the corn, and 86 percent of the cotton farmed in the United States is genetically modified.

Regardless of the cause of increasing food allergy—whether it is semantic or overdiagnosis or related to genetically modified foods—people with allergies still need to be able to shop at the grocery store or eat out at a restaurant. Some general rules can really change their quality of life. First and foremost, avoid exposure to things known to be allergens. A peanut-allergic child should be taught how to avoid foods that clearly contain peanuts.

This first rule is obvious, but the next one is not. In fact, it is a bit counterintuitive. The oil derived from an allergy-producing nut or bean usually does not cause allergic symptoms. This is because the oils don't tend to have the allergy-producing protein. So soy-allergic kids can usually tolerate soy lecithin and soybean oil, and peanut-allergic kids can typically ingest refined or highly processed peanut oil without difficulty.[8]

Another rule is that if the food causes a rash, sometimes cooking the food reduces the reaction. Now it depends on what type of rash we are talking about—total body hives do not count. But if an apple causes a slight red rash around the mouth, applesauce generally won't.

There are also means of predicting whether your child will develop food allergy in the first place. If a first-degree family member has a history of atopy, there is an increased risk of allergy in the child. If a child has eczema in the first six months of life, that child has a greater likelihood of developing food allergy than a child whose eczema appears in the second six months of life.

A lot of parents ask me if coming into contact with allergens increases their child's chance of developing allergies. There are researchers who believe that when a child eats small amounts of a particular food they may build tolerance to the food, but when it is put on their skin, it can cause an allergic reaction. One study showed that if peanut-based oils are use topically—for instance, as massage oils—on newborns up to three months of age, those children have an increased risk of peanut allergy (Strid 2005). But other studies refute this. There is no clear consensus on this one yet. However, I will say that in my practice one of the craziest rashes I ever saw was the result of egg white being used to treat diaper rash. The mom was advised by family to try this home remedy on her slightly atopic two-year-old. Within minutes her child's skin in the diaper area looked like it had a second-degree burn, and the family spent the night in the emergency room. The child was ultimately diagnosed with egg allergy.

As frustrating as it may sound, you cannot just test your child for food allergies. Allergy testing is a wonderful tool, and it can certainly help to predict the likelihood of an allergic reaction. But allergy testing should not be done on everyone. First of all, there are lots of false positives. It turns out that 8.6 percent of the U.S. population tests positive to peanut, while only 0.4 percent have actual clinical allergy with rash, swelling, wheezing, and so on. Looking at the numbers another way, for every hundred Americans with positive skin tests to peanut, fewer than five have clinical peanut allergy (Kim 2008). So we could all rush to the doctor and ask for allergy testing, but the results wouldn't necessarily be helpful—and they may actually add a pretty complicated twist to our lives: thinking we are seriously allergic when we are not.

Another reason that blind testing doesn't work is that allergy testing cannot predict severity. Many allergy tests use a scale of 0–5, with 5 representing the strongest allergic reaction to an allergen. But we pediatricians have all seen food-allergic children tolerate foods to which they tested 5 and have anaphylaxis with others that only registered 2. While this numeric scale is helpful, it has definite limitations.

WHAT IS THE BOTTOM LINE?

Food allergies have increased significantly over the past few years. This represents a real phenomenon—it's not just that we are calling more things "food allergy" but rather that substantially more people are having allergic reactions to foods. While food allergy is an increasing problem, most allergies go away over time. This is more likely with some foods (milk, egg, soy, wheat) than others (peanut, tree nuts, seafood). But ultimately, many kids who have been labeled as food allergic no longer have problems by the time they are in middle school.

There is broad disagreement as to why food allergies have become so common. The theories range from the idea that parents aren't exposing their kids to certain types of food early enough to the exact opposite—that early exposure is the problem. New data suggests that genetic modification of foods is contributing to the rise in allergies, but it is unclear how this works or how big a culprit this new technology is.

If you want to avoid genetically modified foods, the best way to do it is to buy **organic**. Choose foods labeled "**certified organic**" or "**non-GMO**" to maximize the likelihood that you are not exposed to genetic modifiers. This is not a perfect solution, though. Neighboring farms often cross-pollinate, so if one farmer is certified organic but his neighbor plants genetically modified crops, nature (wind, insects, and so on) will naturally mix the genetic pools. While an organic crop is mostly genetically unmodified, it may not be completely pristine.

An alternate way of limiting genetically modified foods in your diet is to limit products containing any ingredients from the food crops that have been genetically engineered: dairy, soy, corn, cottonseed, canola, papaya, zucchini, squash, tomato, potato, beets, carrots, wheat, rice, and so on. This also means avoiding soy lecithin in chocolate, corn syrup in candies, and cottonseed or canola oil in snack foods. Taking this approach is exceptionally restrictive, especially when you consider that the list of genetically modified foods continues to expand over time. I don't recommend it.

Finally, this is one topic where becoming politically active may make a big difference. Genetically modified foods do not require specific labeling. Under current guidelines, the **FDA** does not regulate genetically modified products (like an ear of corn or a potato) because they are whole foods, but a box of cereal is regulated because it is a food product. The FDA has taken the position that genetically modified whole foods are substantially equivalent to unmodified "natural" foods, and therefore they are not subject to FDA regulation. Perhaps consumers should consider demanding that food labels include information about genetic modification so they can make informed choices.

For parents of infants who want to do everything they can to prevent food allergies, you can breast-feed your infant exclusively. This has been shown to decrease the likelihood of eczema and allergy until your child is about two. But then the benefit is gone, and a child predisposed to allergies will become allergic. For parents who cannot, or chose not to, breast-feed, there is no proven benefit of soy formulas or amino acid formulas over cow's milk formulas unless your child is actually allergic to cow's milk. Regardless of what sort of milk you give to the baby, wait to introduce solids until your child is four to six months of age.

If your child is allergic to certain foods, avoidance is important. The Food Allergen Labeling and Consumer Protection Act (2006) mandated clear labeling of the top-eight allergy offenders: milk, egg, soybean, wheat, peanut, tree nuts, fish, and shellfish. A child at risk for anaphylaxis should carry epinephrine in case of emergency. Some people wish to take a "milder" approach and use antihistamines like Benadryl instead of epinephrine. But if a person is showing signs of serious allergic reaction, delaying epinephrine administration could be life threatening. Benadryl doesn't treat anaphylaxis.

WHAT'S IN MY HOME?

My kids are lucky: they have almost no relatives with allergies. So on this front I have been quite liberal. However, I have taken care of a shocking number of peanut-allergic kids with no (or minimal) family history of allergy. After a while, I began to suspect that peanut exposure in the womb or through breast milk may increase the likelihood of allergy. Many studies and the AAP disagree. But I still chose to skip peanuts and peanut butter while pregnant and breast-feeding my own kids, just to put my mind at ease.

Chapter 2

Artificial Sweeteners

WHAT IS THE QUESTION?

For most people, sugar is a treat, a pleasure, a temptation. We save dessert for last, earning it after we eat a healthy meal. Sugar gives us a burst of energy by providing a surge of glucose to rev our body engines.

Of course, we all know that sugar is unhealthy. It causes tooth decay. Like any other high-calorie food, when eaten in excess it causes stomach upset in the short run and obesity in the long run. And it is blamed—probably unrightfully so—for toddlers' uncontrollable behavior and teenagers' pimples.

Ours is a culture of invention: if we want something we can't have, we find a way to have it. So artificial sweeteners were invented, replacing the calories of sugar without reducing any sweet taste. First came saccharin, then aspartame and sucralose and others. With these additives, we were supposed to be able to have our cake and eat it too. Literally.

Artificial sweeteners have become a major component of processed foods over the past twenty years. Each sweetener has been studied thoroughly by the **FDA** and largely deemed safe. But rumors continuously swirl about possible associations with cancer. Are these sugar alternatives really harmless? How about for our children, who will consume these products during their entire lifetime? Or can these sweeteners help in the battle against bulging waistlines? Do the benefits of replacing natural sugar with chemical alternatives outweigh the risks?

WHAT IS THE DATA?

Sugar consumption is an American pastime. The average American ingests twenty teaspoons of sugar every day. About 60 percent of this comes from corn sweeteners and 40 percent from sucrose (table sugar). A very small amount also comes from sweetening sources like honey and molasses, but it is relatively insignificant by comparison.

Sugar has served an important role in our evolution. It is a quick source of energy, a fuel for running, jumping, and other physical activities. In the days of the caveman and the hunter-gatherer, a sugar infusion followed by an energy surge had obvious benefits. But in today's society, which is far more sedentary, sugar often just delivers empty calories.

Sugar itself doesn't make a person obese. Becoming overweight is a result of consuming more calories than your body needs. But sugary foods tend to be high in calories and are often eaten in large quantities. So when a person has too much sugar on a daily basis, the result is a surplus of calories. Excess calories are stored by the body as fat—saving for a day when there may not be enough fuel around.

To lose weight, calorie consumption needs to be decreased or the number of calories burned each day increased. But food that is naturally less caloric usually doesn't taste as good. So dieters often turn to alternative sweeteners with fewer calories, which seems like a reasonable compromise. Ironically, though, as the number of artificial sweeteners on the market has expanded, so have American waistlines. The more people gain, the more "diet" foods they seek out to meet their sweet cravings without fattening them up. Even though this strategy hasn't prevented ongoing weight gain, people continue on their search for a panacea, consuming more and more alternative sweeteners. As a result, over the past 130 years we have seen the introduction of saccharin (first discovered in 1879), aspartame (1965), and sucralose (1976). Newer substitutes include sugar alcohols (sorbitol, xylitol, maltitol, and others) and stevia.

Before 1958 the only sugar substitute available was saccharin, and it was essentially unregulated. Saccharin is three hundred times sweeter than sugar. This potency allows only a small amount of saccharin to be used in order to sweeten a food. Shortly after it was discovered, it became a very effective tool for combating sugar shortages, particularly during both world wars. In 1957 Benjamin Eisenstadt and his son Marvin invented Sweet'N Low, the first marketed and distributed sugar substitute. Soon to follow were the individual-use pink packets still found on restaurant tables today.

In 1958 came the Food Additives Amendment to the Food, Drug, and Cosmetic Act. This required the FDA to provide premarket testing and approval for **food additives**. Because saccharin was already on the market, it was grandfathered out of this requirement. It fell under the category of GRAS, short for "generally recognized as safe," and did not require formal study.

In response to public demand, the FDA began to study many GRAS substances in the early 1970s. In 1972 data emerged suggesting a possible link between saccharin and bladder cancer in rats. This data and its study methods became the subject of much public debate, with claims including that the researchers gave the rats the equivalent of eight hundred diet sodas per day, but saccharin remained largely unregulated. Five years later, in 1977, a Canadian study documented the same association. This time the FDA reacted swiftly by recommending a complete ban on saccharin except as an over-the-counter drug in the form of a tabletop sweetener. Almost simultaneously, Congress passed a moratorium called the Saccharin Study and Labeling Act, and for the next two years, while it was undergoing further safety studies, saccharin was allowed to remain on the market as long as it carried the label "Use of this product may be hazardous to your health. This product contains saccharin, which has been determined to cause cancer in laboratory animals." This moratorium was renewed several times until finally, in 1991, the FDA withdrew its proposal to ban saccharin, citing sufficient safety data in newer studies. In 2000 the **NTP** removed saccharin from its *Report on Carcinogens* and shortly thereafter Congress repealed the requirement for saccharin to carry a warning label.

Even though saccharin was ultimately vindicated, one might ask why a potentially harmful additive would be allowed to remain on the market throughout the 1970s, '80s, and '90s if there was any worrisome data at all. Why was the moratorium continually renewed? To get to the answer, it helps to put saccharin into historical perspective: in 1977 it was the only sugar alternative widely available. So while there were potential health concerns, long-term effects were outweighed by consumer demand. The short-term answer was to keep saccharin on the market with a big, bold warning statement printed on the label. The ultimate solution was to develop a new sugar alternative. Enter aspartame.

Aspartame is an artificial nonsaccharide sweetener sold in little blue packets under the brand names Equal and NutraSweet. It is also used as an ingredient in more than six thousand foods, drinks, vitamin supplements, gums, candies, and other products sold worldwide. Aspartame is 180 times sweeter than sugar. For the same reason as saccharin (its potent sweetness), only a tiny bit needs to be used at any given time. Therefore, it adds sweet flavor with essentially no calories.[1] Initially, the only flaw identified with aspartame was that it degrades with heat, so it is not useful in baking.

Aspartame was discovered in 1965 but faced prolonged FDA approval. In 1981 the FDA sanctioned it for use in some foods but did not consider it universally safe for all foods until 1996. A primary issue at hand was that aspartame is broken down by the body into several chemicals—like methanol, formaldehyde, and formate—each of which is potentially toxic at high doses. However, as the FDA notes on its Web site, citrus juices and tomatoes have similar by-products and, in fact, when metabolized by the body they can contain higher methanol concentrations. In other words, the by-products aspartame creates in the body don't seem to matter much.

There continue to be concerns and claims about links between aspartame and disease. Cancer—particularly brain tumors and lymphoma—remains the biggest. In 1996, just as the FDA was giving aspartame its final blessing, a study was published suggesting that aspartame consumption may be related to an increase in brain tumors (Olney 1996). That study was rebuked by the National Cancer Institute (NCI): a review of its database on cancer showed that the frequency of brain cancers started to increase in 1973—well before aspartame was approved—and after peaking in 1985 has actually decreased slightly. If aspartame were in fact a cause of brain cancer, you would expect to see an increase, not a decrease, in the disease as aspartame use has flourished. Still, NCI continues to study aspartame and currently has a study in progress looking at how dietary factors might be related to adult brain cancer.

As recently as 2006, NCI published data that illustrates no link between aspartame and cancer, and the FDA reaffirmed its vote of safety.

But one year later, a group of European researchers published completely contrary results—their study suggests that aspartame is in fact associated with lymphoma, leukemia, and breast cancer in rats (Soffritti 2007). The seesawing results in aspartame research have become tiresome: Does it cause cancer? Yes. No. Yes. No. Some consumer watchdog groups have upgraded their aspartame rating from "use caution" to "everyone should avoid." The studies are murky enough that it's not worth using the sweetener if you don't have to. But even the executive director of one of the big watchdog groups has said: "The risk to an individual is quite small, so people shouldn't fear that if they have one diet soda a day they are going to develop cancer. . . . [And] if aspartame were that potent a carcinogen, I wonder if we wouldn't be seeing a real epidemic of cancer."[2]

Aspartame is accused of more than just cancer. It is the alleged cause of headaches, dizziness, gastrointestinal symptoms, mood changes, and even seizures. The CDC has explained this by saying that some people may be more sensitive to aspartame than others. The FDA still stands behind its approval of aspartame, though, stating there is no evidence for widespread adverse health effects and that none of the claims about noncancer effects have been substantiated in formal studies.[3]

When the FDA approves a food additive, it must determine a safe dose. In fact, it must identify "the amount of an additive that, if eaten every day for the rest of a person's life, would be considered safe." This is called the **ADI** or "acceptable daily intake." In many cases, the FDA identifies the highest dose that has no effects on experimental animals, divides by one hundred, and that is how much consumers are allowed to ingest. In other words, the ADI represents one one-hundredth the maximal dose found to be safe in lab animals.

For aspartame the ADI is 50 grams per kilogram of body weight. This means that a 165-pound (or 75 kg) adult could safely consume 3,750 mg of aspartame, the equivalent of twenty-one cans of diet soda, daily.[4] A 66-pound (30 kg) child can drink eight cans of soda. The average adult daily aspartame consumption is only about 5 percent of the ADI, or 200 mg per day.

Most aspartame studies equate milligrams of the sweetener with cans of soda. While this is a convenient measuring tool, aspartame is present in many other drinks and foods that we consume every day. Therefore, when you try to estimate your own daily aspartame ingestion, you must consider breakfast cereals, gum, candy, yogurt, Popsicles, and much more, not just soda.

As I am a pediatrician, I must make a brief aside to warn that soda consumption by young children is a health hazard in itself. The more soda a child consumes, the less water or milk she drinks. As a result, soda consumption is often correlated with other poor nutritional indicators. Soda intake is also correlated with an increased risk of obesity. For a child who drinks a soda a day, the risk of becoming obese is increased by 60 percent (Ludwig 2001). Although many discussions of aspartame contain references to soda consumption, this does not mean that the medical community condones drinking soda.

Just as controversy over aspartame was beginning to snowball, along came a different sugar substitute. Sucralose, known by the brand name Splenda (sold for individual use in a little yellow packet), was discovered in 1976 and approved for consumption by the FDA in 1998. Sucralose is six hundred times sweeter than table sugar—that's twice as sweet as saccharin and almost four times as sweet as aspartame. But unlike its predecessors, sucralose is actually made from sugar. Therefore, its manufacturers claim that sucralose tastes more like sugar. It is calorie free, not just because so little is needed in order to sweeten, but also because it is indigestible by the human gut, so it passes straight through your body without being absorbed.

Sucralose has been on the market for ten years and is already used in forty-five hundred different foods and drinks. It seems to be much less controversial than aspartame or saccharin—at least so far. Initial FDA studies found no adverse effects of sucralose. Follow-up studies, though, do show an association between high-dose sucralose ingestion and migraines. There have also been some reports of allergic reactions, and there has been very little data whatsoever about the effects of sucralose on children. But compared with allegations of seizures and cancer, sucralose has fared much better than other artificial sweeteners on the market.

Most rational people would react to all of this controversy over synthesized sugar substitutes by seeking out natural alternatives. Rather than ingest something chemical, why not turn to nature? Genuine sweeteners seem so much more benign.

But the problem with the natural approach goes back to the reason sugar substitutes became popular in the first place: calories. Sugar is natural, as are honey, molasses, evaporated cane juice, rice syrup, and even (pure) apple juice. But these natural alternatives are just as high in calories as sugar, and they tend to stick to teeth, causing tooth decay. There is a new generation of "natural" sugar substitutes called sugar alcohols. They get their name from their chemical composition, but they are neither true sugar nor alcoholic. The sugar alcohols—which include xylitol, sorbitol, maltitol, mannitol, lactitol, and isomalt—occur naturally in fruits and berries. They are lower in calories than sugar and a little less sweet as well, but, unlike all the other artificial sweeteners discussed so far, they are not calorie free.

Sugar alcohols are used most commonly in sugar-free candies, gums, and cookies. When eaten in high concentration, these substances will cause a cooling sensation in the mouth. Hence the refreshing taste of a sugar-free candy or gum. For the same reason, sugar alcohols are used in toothpastes and mouthwashes. Another bonus: sugar alcohols don't cause tooth decay, which is especially beneficial because they are often used as sweeteners in gums.

Sugar alcohols do come with negative side effects, though. The primary complaint is that when eaten in excess, these ingredients can cause either bloating or diarrhea. Like the natural sugars in fruit juice, too much has a laxative effect. The other complaint has to do with weight gain: when people overindulge, sugar alcohols cause weight gain. This is simply a reflection of the fact that these sweeteners are not calorie free.

The FDA considers the safety of each sugar alcohol individually. Some fall under the GRAS provision, as they were used widely before the Food Additives Amendment to the Food, Drug, and Cosmetic Act in 1958. Remember that GRAS just means "generally recognized as safe." This classification comes from the rationale that if something has been in widespread use for many years and has no correlated dangers, or if qualified experts have deemed it safe, then it must be fine. Of course, the flip side of that argument is that a GRAS substance has yet to be studied carefully because it is not subject to the level of scrutiny otherwise imposed by the FDA and other regulatory agencies.

Sugar alcohols seem to be following a trajectory similar to sucralose:

while some complaints have been raised, there has been no controversy approaching the drama surrounding saccharin or aspartame.

The last on the list of sugar alternatives is stevia, a sweetener derived from a South American shrub. Stevia extracts are up to three hundred times sweeter than sugar, offering a familiar low-calorie alternative. Stevia has been a ping-pong ball in an FDA labeling game, resulting in a complex series of reclassifications. Now to set the stage, you should know that the FDA defines a food additive as a substance that will become part of a food and in many cases will change that food's consistency, flavor, or packaging. A dietary supplement is quite different: it is a product meant to augment the diet by adding a vitamin or mineral or even medicine.[6]

Between 1991 and 1995 stevia was actually considered an "unsafe food additive." It was upgraded to "dietary supplement" in 1995. Even during the first printing of this book, neither it nor any of its derivative forms could be sold under the title "sweetener" or "food additive," and stevia itself is still sold as a "dietary supplement." But in late 2009 the FDA granted "no objection" status to the use of a stevia extract as a "food additive." This extract is called by many names, including Rebaudioside A, Rebiana, and Reb-A (also known by the brand names Truvia and PureVia). This further adds to the labeling confusion about this family of sweeteners. In short, only one extract of stevia is FDA approved and the rest of the family—including sweeteners made from the whole leaf—are not.

All of this controversy over stevia's labeling came from questions about its safety. There were published studies implicating it as a mutagen, a substance that causes cells to mutate and become precursors to some cancers. As a result, stevia was denied classification as a food additive and instead called a dietary supplement. Why does this matter? Well, for people in the business of selling stevia the classification as a dietary supplement significantly reduces its presence in processed foods. The gum or candy or breakfast cereal you buy at the store is sweetened with a food additive, not a dietary supplement. But it seems pretty clear that stevia is not a supplement—it is not meant to be a vitamin or added bonus to your daily intake, and it is certainly not a medicine. It is meant to replace sugar. So essentially, the FDA has limited stevia's ability to break into the processed foods market by calling it something (a dietary supplement) that it isn't.

While it may sound like I am on stevia's side, I am actually not. Rather, I am bothered by the FDA's approach. Instead of rejecting stevia as a sweetener altogether, the FDA essentially gave it a demotion and called it a dietary supplement. There is published data showing safety concerns and possibly even data demonstrating true risks of ingesting stevia. Why, then, is stevia allowed to be sold as anything? If there are real concerns, why should stevia appear on the shelf at GNC with multivitamins and immune boosters? Rather, if stevia—or any product—is potentially dangerous, I think it should be banned from the marketplace altogether.

Consumers don't recognize the semantic twist of swapping "supplement" for "additive"—and why should they? In the case of stevia, the labeling has been sneaky. By definition, the FDA does not guarantee the safety of dietary supplements. Presumably some—like stevia—aren't safe enough to be called food additives. But nutritional supplements and vitamins have become massive players on the supermarket shelves and in their own storefronts. These products continue to be peddled to the public without concern. The example of stevia makes me wonder what health claims exist against other dietary supplements.[7]

Perhaps all of this is overkill. Stevia extracts are used around the world, including Japan and many countries in South America. In 2006 **WHO** published a formal statement that stevia and its breakdown products are not mutagens and there is no evidence linking them to cancer. Furthermore, WHO suggested that there may even be health benefits to stevia for certain patients, like those with hypertension or type 2 diabetes. These health claims have surfaced many times over the past few years, but this was the first time a major health organization invoked them in a policy statement.

WHAT IS THE BOTTOM LINE?

There is a tremendous amount of conflicting data about sugar substitutes. Americans are always looking for a calorie-free way to eat sweets. The problem is that in order to avoid sugar calories our treats have to be sweetened with artificial sweeteners. While there is no hard-and-fast data showing a link between saccharin or aspartame or sucralose or stevia and disease in humans, there is suspicion fueled by animal studies.

To complicate the issue, future generations of children (unlike today's adults) will have a lifetime of exposure to these sweeteners, and we have no idea what the long-term implications are. The other problem is that even with calorie-free sugar alternatives, Americans continue to gain weight at a rapid clip. Portion size in this country is enormous, and exercise is at an all-time low. People think that the words "diet" or "sugar free" on a label mean they can eat as much as they want and not gain weight. This is far from the truth, and there are major long-term health consequences to being overweight or obese.

So ultimately, this is yet another area where the general rule of thumb should be: everything in moderation. An occasional diet food is fine, but stocking the pantry with them may not be. We should go back to the philosophy that sweets should be consumed less frequently. It looks like a home-baked cookie made with real sugar is going to turn out to be a whole lot healthier than high-volume consumption of artificially sweetened treats claiming to offer less fat, less sugar, and more taste.

WHAT'S IN MY HOUSE?

We have regular old sugar (white, light brown, and dark brown) and honey in our house. My husband bought a box of Sweet'N Low about five years ago just to have around for guests, but it sits in our pantry untouched. We don't keep sodas at home. I am not as strict as I sound—in fact, my kids are allowed a sweet a day, and I have been known to break that rule and give more. But I would much prefer that treat to be ice cream or a homemade brownie rather than a soda or a candy bar. My kids have no taste for "diet" foods and I hope it stays that way.

Chapter 3

Baby Foods

WHAT IS THE QUESTION?

Around six months of age, a baby is ready to start eating solid foods. Parents with the time and the inclination can make baby foods at home. Fruits and vegetables may be steamed, pureed, and then eaten fresh or frozen into ice cubes for storage in the freezer. But many parents work long hours or don't have easy access to fresh produce. For them, prepackaged baby foods are the only alternative.

Along with feeding your child premade foods comes a lot of guilt. This didn't used to be the case. A generation ago, prepared foods were a welcome relief for parents. These baby foods might not have looked very appetizing, but they were largely considered safe and healthy. Today every consumer item is scrutinized, down to the sixty-nine-cent can of pear puree.

So is it okay to give your child canned food off the shelf? If you do, does it have to be labeled "**organic**"? Are refrigerated purees worth their higher price tags and actually healthier for your child? How can parents navigate the booming industry of baby foods?

WHAT IS THE DATA?

Babies are on all-milk diets (breast milk, formula, or a combination) from birth until they are about five or six months old. Older infants need more calories and are ready to explore the texture of food. These babies are more social beings who want to sit at the table and eat, and who have also learned how to put just about everything within reach into their mouths. When your baby stares at your food, tries to grab it, puts it in his mouth, or smacks his lips at the sight of your lunch, he is begging to start solids.

Of course, six-month-old babies are not ready to chew or manage big bites. They gag on any kind of chunk—even a fork-mashed banana can be too much to handle. This is because in the early stages of eating a baby is learning to take food off a spoon, manipulate it with his tongue and gums, and then swallow. This sequence of events takes time to master, and that is why infants should have only purees for the first two to four months of eating solid food.

Fruits and vegetables tend to be recommended starter foods. Some pediatricians and nutritionists even have a preferred order. The rationale here is that most produce (except for carrots, spinach, berries, and citrus) tends to be well tolerated and does not trigger allergic reactions. Of course, this may vary from baby to baby. Rice cereal is another staple starter food. It offers a grainy texture but is very easy for most babies to manage. Rice cereal also adds calories to the meal and comes fortified with iron, both necessities for many rapidly growing babies.

It is pretty easy to make starter foods at home: if you have the time, the equipment (a blender or a pureeing device), and access to produce, that is. I won't go into the mechanics because there are several good books in print with recipes and techniques for making infant foods.

But many parents lack the time or equipment or access to ingredients. Personally, I couldn't make homemade purees for my babies while trying to juggle a full-time medical practice and the basic necessities of life—I could barely fit in time to pay the bills. If you are like me and cannot do it all, you know the dilemma of walking into your local supermarket and being confronted with an aisle—sometimes fifty feet long—with shelves full of different baby foods. How do you decide which to feed your child?

There is actually a rational way to approach this seemingly endless array, because there are a few distinguishing features among baby foods. First, there are jars labeled "organic" and jars that aren't. Then there are jars that contain preservatives and others that don't. Finally, there are jars that remain at room temperature and refrigerated purees.

Certified organic baby food purees come from produce grown without pesticides, fertilizers, antibiotics, hormones, genetic modification, additives, irradiation, or sewage sludge. It's pretty obvious that if you can avoid giving your baby these chemicals and additives, you should. Some researchers believe that babies are at higher risk from pesticides and hormones because their developing bodies manage these ingredients differently than yours and mine do. Many experts say that a baby's dose of these additives is much higher—because if a child is eating a pureed, concentrated vegetable mush that has pesticide residue, the amount of pesticide per unit of body weight is going to be higher than the amount taken in by an adult who is eating the whole vegetable (and for whom the vegetable isn't the entire meal).

These arguments may turn out to be true or they may not, but it doesn't really matter. We all intuitively know that a food with no hormones, no pesticide residues, no modified genes, and no antibiotics is better than the alternative. The major stumbling block for many parents, though, is cost.

WHAT DOES THE WORD "ORGANIC" MEAN?

When "organic" appears on the label, it can mean a number of different things. According to the USDA and FDA labeling requirements, these are the variations on "organic":

"100 percent organic": Must contain 100 percent organically produced ingredients, not counting added water and salt.

"Organic": Must contain at least 95 percent organic ingredients, not counting added water and salt; must not contain added sulfites; may contain up to 5 percent of (a) nonorganically produced agricultural ingredients that are not commercially available in organic form and/or (b) other substances allowed by 7CFR205.605.[1]

"Made with organic ingredients": Must contain at least 70 percent organic ingredients, not counting added water and salt; must not contain added sulfites (except wine, which may contain added sulfur dioxide); may contain up to 30 percent of (a) nonorganically produced agricultural ingredients and/or (b) other substances, including yeast, allowed by 7CFR205.605.

Organic claim on the information panel only: May contain less than 70 percent organic ingredients, not counting added water and salt; may contain over 30 percent of (a) nonorganically produced agricultural ingredients and/or (b) other substances without being limited to those in 7CFR205.605.

Until recently, organic foods were in the strict minority. This has changed since the new millennium, and these days organic farms have a much larger share of the marketplace. But these farmers are still far less common than their nonorganic counterparts and therefore face higher overhead costs. The cost is passed along to the consumer, so a jar of baby food can be significantly more expensive if it carries the word "organic" on its label.

Is it worth the extra cash? Probably. We really don't know what most pesticides, hormones, and other unnatural additives do to growing bodies and minds. And because we don't know, avoidance is a good strategy. But do the best you can. If your local grocery store doesn't offer organic baby foods, you may have to choose a nonorganic variety or order them from the Internet instead. If you cannot afford organic baby food every week, buy it when you are able and otherwise use nonorganic.

How about preservatives? A preservative is anything that extends the shelf life of a food. We think of these substances as added chemicals, but that is not necessarily the case. Yes, baby foods that are sold on supermarket shelves need to have some form of preservative. However, there are lots of nonchemical options.

For instance, a common form of preservation is canning: storage in a jar, foil, or plastic container, sometimes even a box. When canning was first invented, around 1825, chemical preservatives weren't available. Canning increases the longevity of food because the food is boiled first in order to kill any bacteria—if the contents are completely sterile, the food does not spoil. Sometimes the food is sealed into the container before it is actually boiled; sometimes the sealing occurs during the boiling process. Either way, after you get rid of all microorganisms the food cannot rot or mold. If you open it, though, new bacteria can come in—hence the "refrigerate the contents after opening" warning on most cans and jars.

Refrigeration is another means of preservation. By keeping food cool, bacteria cannot thrive nearly as well as they can at room temperature. Eventually molds do grow, but it takes much longer in a fridge. Freezing is even more successful than refrigeration because at freezing temperatures bacteria are completely inhibited. Food in a freezer may get freezer burn, but it will never rot.

Chemical preservatives are also used to extend the shelf life of foods.

The most common include benzoates, nitrites, sulfites, and sorbic acid. All of these preservatives work the same way: they either kill bacteria or make it impossible for bacteria to grow and multiply. There is mixed data on these chemical preservatives. The **FDA** finds them safe enough to include in store-bought foods, but there are plenty of concerned consumers out there. Sodium benzoate is generally considered safe but has been accused of causing allergic reactions, exacerbating attention problems, and even causing cancers.[2] Nitrates, which are found mostly in meats (because they help maintain color and flavor), can be precursors to the cancer-causing chemical nitrosamine, and as a result this particular preservative has been gradually phased out over the past few decades. Sulfites can cause problems for people who are sensitive to them—some asthmatics wheeze (occasionally severely) and some migraine sufferers experience headaches.

Just because a jar of baby food is "preserved" in order to be able to last on the shelf doesn't mean that it is chock-full of chemical preservatives. And even if it does have some additives that help with preservation, for the vast majority of people these additives are fine.

Most baby foods use no added preservatives. Commercial baby food manufacturers—like Gerber, Beech-Nut, and Earth's Best—are required by law to list the ingredients contained in each jar. As a result of growing pressure to remove many additives, few prepackaged baby foods now have anything extra added: no salt, sugar, spices, or preservatives.

Baby foods requiring refrigeration may appear "fresher" because we associate keeping things in the fridge with freshness. But this doesn't mean that it is better for your baby—refrigeration is simply being used as a form of preservation. If the refrigerated foods are two or three times the cost of room-temperature foods in jars with no added preservatives, you can feel comfortable spending less because the refrigerated variety is not necessarily any healthier.

One last note about baby foods: in general, homemade baby food has more calories per spoonful than prepackaged foods. This isn't always the case, but it happens much of the time. The reason is that homemade food is typically thicker, with less added water. So if your baby is gaining weight slowly, or spits up easily, it may be worthwhile to consider making your own food. For the skinny baby, more calories per spoonful can translate into better weight gain; for the "spitty" baby, more concentrated foods may be eaten in smaller volumes and therefore may stay down better.

WHAT IS THE BOTTOM LINE?

Homemade baby foods are ideal. But if you don't have time to make them or the equipment or access to fresh fruits and vegetables, don't lose sleep. There are healthy and relatively inexpensive alternatives available.

Look for foods in jars with no added preservatives. If available, choose organic baby foods because they should contain fewer pesticide residues, hormones, antibiotics, and genetically modified products. But again, if organic baby foods aren't accessible or if they are too expensive, it is okay to give your child nonorganic ones. Finally, while refrigerated baby foods may seem fresher, these foods aren't always healthier for your child. Room-temperature foods in jars with no additives may be available at a lower cost, and they are equally good.

The most important aspect of introducing a baby to solid foods is really the socialization: meals should be healthy and fun. Fresh fruits and vegetables will, hopefully, become a staple of your child's diet. Sitting for meals and talking and laughing with your child are all integral parts of the experience. While the quality of the food is clearly important, it is only one ingredient in healthful eating.

WHAT'S IN MY HOUSE?

When they were babies, my kids ate purees from jars. My husband and I were both working around the clock, so there just wasn't free time to make baby food from scratch. I bought organic foods as much as possible. Sometimes my husband or I would try to fork-mash a banana or avocado, but our son proved to have a superhuman gag reflex and couldn't tolerate the smallest chunk until he was about ten months old.

Chapter 4

Fish

WHAT IS THE QUESTION?

Fish was once considered the "healthy" protein. Lower in fat than many meats and dairy products, it was a darling of the American diet for decades. Fish could be a tough sell for children, but parents worked enthusiastically to get it in, succeeding most often in the form of tuna fish salad or fish sticks.

Then, in the mid-1990s, reports about mercury contamination of fish began to emerge. In 2004 the story jumped into the headlines when the **FDA** issued a well-publicized fish-consumption advisory. All of a sudden, fish received a ton of bad press and people were very confused. Tuna went quickly out of vogue (as did the notorious stars of the advisory: swordfish, tilefish, shark, and king mackerel). But people were urged to keep eating at least some fish, because their omega fatty acids were more than healthy—they were downright life extending.

Tuna was out and salmon was in, poised to take over as America's favorite fish. But no sooner had it moved to center stage than salmon too got its fair share of criticism. Wild salmon was fine, but the cheaper, more readily available farmed variety was not. In fact, it too was labeled as dangerous.

Today we are left in a puddle of confusion. If fish is potentially tainted—with mercury or toxins or colorings or whatever—shouldn't we just skip it altogether? How important are the omega fatty acids? And can you get the benefits of the omegas without actually eating fish? Can you find a fish oil supplement that doesn't have all this bad stuff in it?

WHAT IS THE DATA?

Until just around the turn of the millennium, fish was considered a fabulous food. It is generally low in fat, provides a high protein load, can be served raw or cooked, and is widely available all over the world.

Fish is also chock-full of omega fatty acids, particularly omega-3.[1] These omega fatty acids were originally called "essential fatty acids" because they cannot be manufactured by the body. Fatty acids were initially thought to be critical for growth in young children, but now we know that many of them benefit people of all ages. In the past decade, omega-3s have become the celebrities within the group. Two particular omega-3s—eicosapentaenoic acid (**EPA**) and docosahexaenoic acid (**DHA**)—are credited with reducing inflammation and helping to prevent chronic illnesses like heart disease, arthritis, and certain types of cancer.

Fish was hailed as the ultimate food until 2004. Before then, there were rumblings that fish was tainted with mercury, but this notion didn't make it into the mainstream. The Environmental Protection Agency (EPA) and the National Academy of Sciences had released statements in the 1990s, but none of them gained much traction. I had a handful of patients whose chiropractors or acupuncturists recommended mercury testing of the hair and nails, but it was pretty much a fringe element. Then, in 2004, the FDA published recommendations that some people should significantly decrease their fish consumption because of mercury contamination in the ocean. Fish went from beloved to feared, and the concern over mercury wasn't on the fringe anymore.

The FDA recommendations weren't for everyone; they were only for women of childbearing age and children. But the message was stated in blunt, well-publicized English: these groups should not eat shark, swordfish, king mackerel, or tilefish because they are all high in mercury. According to the FDA, the risks of mercury contamination to the developing brain and nervous system are so great that they offset the benefits of this omega-rich food. People listened, maybe too well. Though the statement went on to say that these women and children

---

2004 FDA RECOMMENDATIONS ABOUT FISH CONSUMPTION FOR WOMEN OF CHILDBEARING AGE AND CHILDREN

1. No shark, swordfish, king mackerel, or tilefish (all high in mercury).
2. Do eat up to twelve ounces a week (average serving size: six ounces) of low-mercury fish (e.g., shrimp; light canned tuna, which is lower in mercury than albacore "white" tuna; salmon; pollack; catfish).
3. If you are going to eat local/fresh fish caught by friends, check local advisories, and if there is no info available, only eat six ounces a week.

Note: The FDA's maximum allowable concentration of mercury is 1 ppm (part per million).

---

can, and probably should, eat up to twelve ounces per week (two servings) of other low-mercury fish, many decided to stay away from fish altogether.

Mercury wasn't a new player. It is a naturally occurring element that has been around for millions of years. But these days, mercury appears in landfills where consumer goods are dumped, ultimately seeping out into nearby soil. It is also a by-product of manufacturing waste, released into the air through industrial pollution. Aerosolized mercury ultimately falls to the ground and accumulates in streams and oceans.[2]

In the water, bacteria transform mercury into methylmercury. Fish then accumulate the methylmercury in two ways: first, as water passes over their gills methylmercury is absorbed; and, second, when fish feed on smaller fish, they ingest the mercury that the littler species have previously ingested. Once consumed, mercury tends to stay in the fishes' bodies. This is called bioaccumulation. Because mercury is deposited in all our oceans, lakes, and streams, all fish (and shellfish too) have some methylmercury. If you think about it, it is logical that bigger fish have higher mercury contents than smaller fish. The bigger fish eat the smaller fish, accumulating the mercury stores of their prey. And these larger fish tend to live longer, so, as a function of time, they are also exposed to more mercury-contaminated water passing over their gills.

But all of this evidence still doesn't explain why the FDA made a blanket statement in 2004 about fish consumption specifically for women of childbearing age and children. Mercury has been around a long time—it was on the periodic table when I was in grade school and no one seemed to lose sleep over it. In fact, it provided endless amusement for kids of my generation when we realized that we could break open a thermometer and play with the gelatinous mercury that oozed out.

As it turns out, mercury has been a known troublemaker for centuries. The phrase "mad as a hatter" actually comes from the physical symptoms—including trembling (sometimes called hatter's shakes), loss of coordination, slurred speech, irritability, loss of memory, and depression—associated with mercury nitrate used by felters who manufactured hats. In the 1940s and 1950s, when some babies were exposed to mercury in teething powders and other products, they developed acrodynia, or pink disease.[3]

The issue in 2004 was recognition—or at least public awareness—of the fact that even small doses of methylmercury can be toxic to some nerves in the body, including those in the brain. When fish accumulate methylmercury, it is tightly bound to proteins like muscle. You cannot cook mercury out of fish—sautéing or broiling a piece of fish does not reduce its methylmercury content. When we eat fish, we absorb almost 100 percent of its mercury. This is because methylmercury can traverse pretty much every cell wall in our body. It has a particular affinity for our brain and muscle cells.

In 2004 it was publicly recognized that methylmercury can—at least potentially—permanently affect the development of the brain and nervous system. Government agencies began to acknowledge that methylmercury crosses the placenta easily and that fetuses hold on to mercury more effectively than adults.[4] In fact, if you measure the mercury level in a baby just after birth, it can be 30 to 100 percent higher than his mother's (Sakamoto 2004). So if the developing brain is vulnerable to mercury, and fetuses are not only exposed to it in the womb but actually hold on to it steadfastly, there is a chance that fetal brain and nervous system development will be (and is) affected by this compound.

That is why the FDA statement focused on women of childbearing age and children. It is fairly obvious that children with rapidly developing brains and nervous systems should steer clear of mercury. But during the nine months when they are in the womb and often for the first few months afterward, babies rely upon their mothers as sources of food. Therefore, a pregnant woman should avoid exposing her fetus to mercury. Mercury is excreted in the breast milk of lactating women, so breast-feeding women are counseled to keep a low-mercury diet as well.

Even if a woman is not pregnant or breast-feeding, remember that mercury likes to bioaccumulate—it takes somewhere between fifty-four and seventy days for the body to get rid of just half the mercury it has consumed (this is called its "half-life"). So if a woman wants to become pregnant, it can take months (some say years) to get mercury down to an acceptably low level.

With its 2004 advisory, the FDA was following in the footsteps of the EPA and the National Academy of Sciences, both of which recommended lower fish consumption in order to reduce mercury exposure back in the 1990s. The FDA decided that fish with more than one part per million (ppm) of mercury—specifically shark, swordfish, tilefish, and king mackerel—are the ones that women and children should not eat. One ppm means one milligram per kilogram of body weight—of the fish, that is. Fresh and frozen (particularly albacore) tuna are also included on many lists of fish with high levels of mercury because their mercury content can vary from a low of 0.5 ppm to a high of 1.5 ppm. Canned "chunk light" tuna is not considered worrisome because its mercury content is consistently low.[5]

Following the 2004 FDA statement and its accompanying media blitz, people became concerned that they had already ingested massive amounts of mercury. Sushi connoisseurs, especially, panicked about their health-conscious eating choices being harmful. Many people wanted to know what their mercury level was. The two most popular ways to measure this look at hair samples or blood samples. Though many patients rushed to doctors to ask for these tests, most doctors hesitated because they didn't know what the results meant. None of us should have mercury in our blood or hair, so a normal level is zero. But almost all of us have at least some, because we live in a mercury-contaminated world.[6] Today it is widely accepted that any person—child, woman of childbearing age, adult male, whoever—should be aware of the mercury content of fish and should probably consume a low-mercury diet.

As for the toxicity of mercury on the developing nervous system, we know that high doses of mercury can have profound effects including movement abnormalities, spasticity, and seizures. But low-dose mercury exposure is the real issue for most fetuses, and we still do not know what it does. General brain development—including cognitive thinking, memory, attention, language, and motor skills—are all thought to be vulnerable to methylmercury. But the actual neurological effects (if there are any) at low doses are not yet known. This is how the FDA rationalizes singling out women of childbearing age and children in their warning.

The concerns about methylmercury are legitimate, but these days mercury has a mighty bad reputation. It might as well be a four-letter word. Here's where you need to be careful, because there are different types of mercury. Methylmercury—the form found in seafood—is not the same as ethylmercury, which was once used as a preservative in childhood vaccines and is often known by the name thimerosal. Both are organic forms of mercury that cross cell membranes easily, including the blood-brain barrier that normally protects our brain from infections and toxins. But methylmercury is a by-product of bacterial conversion of mercury in oceans and streams. It can stay around in the body for months or years. There are serious concerns about methylmercury and the developing brain because methylmercury bioaccumulates, crosses the placenta, and seems to have the potential to damage neurons. On the other hand, ethylmercury leaves the body quite rapidly (in a matter of days). And study after study fails to prove any connection between ethylmercury and developmental disorders of the brain. So when you hear something about "mercury," consider the type before coming to any conclusions.

Mercury was just the beginning of trouble for fish. The FDA recommendations came out when my daughter was a year old, just about the time she was starting to eat adult food. I read all the data about mercury, but I also knew the benefits of fish in a child's diet. We decided early on to introduce her to salmon as an alternative to high-mercury fish, and she loved it.

Not long after, I was racing through the supermarket in an effort to do the grocery shopping on my way home from work. I was already late but had promised to pick up what we needed for dinner. I ran up to the fish counter to get a small salmon filet. I knew I was supposed to buy wild salmon but the only options available were Atlantic or Pacific. Was one of these code for "wild"? One looked pinker than the other. That's good, right? It's the color that salmon is supposed to be. Or did it mean that it had chemicals added to enhance its color? My quick trip suddenly turned into something a lot longer.

The debate about wild versus farm-raised salmon actually started before all the drama about mercury; it just didn't come to center stage. Salmon first fell into the spotlight in 2001, when the BBC produced a report on the subject. Two years later, the Environmental Working Group (EWG) released a study stating that farm-raised salmon from the United States had the highest **PCB**[7] levels anywhere in the world. And they were certainly higher than the PCB levels in other U.S. fish—on the order of 16 times higher than wild salmon, 3.4 times higher than shellfish, even 4 times higher than beef. Then, in 2004, a study published in the journal *Science* said that farm-raised salmon had 10 times more toxins than wild salmon. The authors cautioned consumers to eat farm-raised salmon no more than once every one to two months.

So what's so bad about farm-raised salmon? For starters, salmon raised on farms are fattier fish. This means that they have more fat cells, and fat cells are where PCBs and closely related **PBDEs** are stored. As a result, they tend to have higher levels of these industrial chemicals. Also, because they have more fat they have less protein per serving. And most farm-raised salmon are fed fish-meal pellets made from ground-up sardines, anchovies, mackerel, and other small fish. The pellets also contain high levels of PCBs, antibiotics, and other unsavory by-products. This isn't how salmon would feed in the wild, and the effect is to further increase their load of toxins.

Add to this that the fatty acids that make all fish—particularly salmon—so beneficial can be grossly out of balance in farm-raised salmon. The family of omega fatty acids can be broken down into two groups. The omega-3s are essentially the "good" omegas, hailed for improving inflammatory conditions (like arthritis) and mood disorders (like depression) and possibly even reducing the risk of heart disease. Omega-6s, on the other hand, seem to do the opposite. They do play some beneficial roles, but they also can increase inflammation and depression. As it turns out, farm-raised salmon have more omega-6s than wild salmon. Even though the total number of omega-3s may be similar in wild and farmed salmon, omega-6s compete with the 3s in our body. So the net effect is that when you eat a farm-raised salmon, you get relatively fewer omega-3s and more omega-6s.

Then there's the issue of how various salmon look. Farm-raised salmon tend to be grayer than their wild counterparts. In nature, salmon's pink color comes from canthaxanthin, a naturally occurring coloring agent from the family of carotenoids that make carrots orange. Canthaxanthin is also what gives shrimp their pinkish orange stripes. In fact, wild salmon look pink specifically because they eat fish—such as shrimp—that have naturally occurring canthaxanthin. But canthaxanthin does not occur naturally in salmon, so farm-raised salmon feeding on fish pellets rather than wild shrimp do not accumulate the pink-orange pigment. In order to make farm-raised salmon look more appealing to the consumer, they are often given supplemental colorings like synthetic canthaxanthin. This synthetic coloring is not flagrantly dangerous when consumed, but it would be nice to avoid any additives—especially food colorings—in foods that are presumed to be fresh and natural.

There is actually a fairly long list of other chemicals and additives found in farmed salmon: dioxins, pesticides, antibiotics, and copper sulfate, to name a few. Many of these substances show up in wild salmon too, but their levels in farmed salmon are much higher.

How do you know what kind of salmon you are buying? It turns out that Pacific salmon is usually wild, while Atlantic salmon is almost always farmed. In fact, farmed Atlantic salmon accounts for approximately 85 percent of all of the farmed salmon in the world. There is no such thing as wild Atlantic salmon available—there are so few of them left that the species is endangered. So it turns out that the labels on fish—Atlantic and Pacific—are in fact code words.[8]

Another twist in the labeling of salmon: if you think you are buying wild—I mean, it says so on the label, for God's sake—wild salmon aren't always truly wild. For one thing, many spend the first half of their lives in hatcheries. While these fish live out their mature days in the wild, they still spend a significant amount of time being exposed to all the PCBs, antibiotics, fish-meal pellets, and other negatives of the farm. To make things even more misleading, sometimes wild salmon are simply mislabeled: they aren't wild at all. Wild salmon are seasonal fish, caught between May and October. If you are buying "fresh wild salmon" (that hasn't been frozen) in the middle of December, something's not right.

Even though it seems pretty obvious that wild salmon is the better choice, you don't have to make yourself crazy over this. Yes, farmed salmon have higher levels of PCBs, dioxins, pesticides, and other toxins than most wild salmon. But several studies (along with the FDA) argue that the levels of all these chemicals are so low in both types of fish that consumers don't need to worry much about it.

All the focus on the potential hazards of fish—the mercury, the PCBs, the by-products of eating a farmed fish—has created a new, perhaps more serious issue: people aren't eating as much of it anymore. By avoiding fish, it is true that you are reducing your exposure to potential toxins. But you are also missing important omega-3s. In 2005 a study out of Harvard concluded almost the opposite of the 2004 FDA recommendations. It found that diets rich in seafood are extremely beneficial for a child's brain development. Moreover, children of women who during pregnancy ate a variety of seafood *specifically including canned tuna* had higher vocabulary test scores compared to children of women who reported avoiding fish during pregnancy (Oken 2005). Could this issue get any murkier?

There are still people who want to stay away from fish, and this is understandable after all the bad press that has circulated. But the omega fatty acids, especially the omega-3s, are important. For these folks, I recommend fish oil supplements. I used to be a skeptic about fish oil, certain that it was high in mercury and PCBs just like the fish the oils are derived from. But methylmercury really accumulates in the meat of the fish, not in its oils. I was pretty surprised to learn that fish oil supplements—before they are even "treated" to remove any heavy metals—contain almost no mercury.

Now the FDA does not strictly regulate vitamins, herbs, or nutritional supplements, which means that the government doesn't guarantee their strength, purity, or safety. This can get tricky for a person who is very conscientious about avoiding mercury. The label of a fish oil supplement may say "mercury free," but the FDA hasn't necessarily checked it. While it is true that most fish oils have much lower PCB, methylmercury, and pesticide concentrations than fish meat, this is not necessarily true for unrefined fish oils. So if you purchase fish oils, make sure they aren't unrefined. If you choose supplements stamped with quality seals from the NSF or the Natural Products Association (formerly NNFA), you are guaranteed the product has no, or trivially low, levels of toxins or heavy metals.

Some people choose to stay away from fish altogether: they don't want to eat the meat and they don't want to ingest the oils. For this group, an omega fatty acid supplement in the form of flaxseed, flaxseed oil, walnuts, or hemp seed can be very beneficial. Each of these has alpha-linolenic acid (**ALA**), a precursor to the other omega-3s. In fact, the body can make all the DHA and EPA it needs if you just give it ALA. The trick is that you need to give it enough.

Today many obstetricians recommend that pregnant and breast-feeding women take omega-3 fatty acid supplements. Omega-3s (particularly DHA) are believed to help fetal growth by increasing birth weight and length. Fatty acids are considered so beneficial to infant growth and development that now they are even added to most infant formulas. You will see them advertised as DHA, ARA (short for arachidonic acid), or some other omega fatty acid additive. If you are buying the "advance" or "lipil" formulas, you are getting formula that is fortified with these fatty acids.

There is such a thing as too much of a good thing. If you give your child (or yourself) too much omega-3 on a regular basis, your child (or you) could begin to have problems with bleeding. Even though this is a "natural" supplement, you are essentially using it like a medication. Excessively high doses of anything can be dangerous.

WHAT IS THE BOTTOM LINE?

So now what do we do? The FDA has told us to avoid giving our kids fish containing high levels of mercury because methylmercury can damage the developing brain and nervous system. We have even become skittish about tuna—a cheap and readily accessible staple in the majority of homes until a decade ago—because we are told that sometimes it has high levels of mercury.

Many people turned to salmon in order to maintain their omega-3 intake, only to learn that the majority of the salmon on the market is farmed. This farmed salmon has less omega-3 benefit and carries the burden of PCBs, antibiotics, and other farm residue.

Just as it seemed like the right thing to do might be to skip fish altogether, data emerged that a fish-containing diet directly benefits the brain of a child. In fact, it seems as though babies and kids really need fish. So now we are back to square one.

My bottom line is this: everything in moderation. It's just logical. You don't want to give your child lots of methylmercury. On the other hand, you don't want to deprive the growing brain and nervous system of omega-3 fatty acids.

Because of industrial waste, we know for a fact that fish contain much more methylmercury today than when we were children. There are many folks working hard to change industrial pollution standards, and this will clearly benefit future generations—not necessarily our children but almost certainly theirs. You can join the movement by becoming involved with the EPA or your local representatives. But for now, our children eat seafood with methylmercury, whether we like it or not.

You could stop eating fish altogether, though I don't think this is necessary or wise. The omega-3 fatty acids are critically important, and you need to consume them. If you do choose to take a break from fish, fish oil supplements and even flaxseed, flaxseed oil, walnuts, and hemp seed are reasonable alternatives. When you purchase fish oils, just make sure they are not unrefined and look for an NSF or Natural Products Association seal of approval on the label—this is as close as you will get to a guarantee that they are heavy metal and toxin free.

WHAT'S IN MY HOUSE?

I continue to give my kids fish. They eat salmon about once a week. I do my best to buy wild salmon, though I recognize that I don't always know for sure if I am actually getting what I think I am. My kids don't have a taste for tuna, but I still do. Tuna salad made from canned chunk light tuna continues to be part of my diet about once or twice a month.

TEACHING KIDS HOW TO CHOOSE SAFE FISH

The EPA has a section on its Web site dedicated to teaching children about healthy fish choices. Fish Kids (www.epa.gov/fishadvisories/kids/) is an interactive learning site with games and simple teaching tools.

Chapter 5

Probiotics

WHAT IS THE QUESTION?

Humans are not solitary beings. We depend upon all sorts of tiny organisms to help our bodies function maximally. Yes, bacteria, viruses, and parasites live on and in each of us. These "bugs" need us to provide food and shelter, and we need them to perform important biological functions. This codependency is probably best illustrated in our intestine. Here, bacteria line the walls, aiding in the digestion of food and helping to produce important chemicals such as vitamin K. Many people refer to these bacteria as "good bugs."

When your child is sick and takes an antibiotic, the medicine wipes out the bacteria responsible for the illness. But that antibiotic kills many of the good bacteria in the body at the same time. Probiotics are essentially the opposite of antibiotics. These are good bugs that can be taken just like medicine. In theory, at least, probiotics replenish the supply of the helpful microorganisms, aiding and abetting various parts of the body from the intestinal tract to the bladder to the immune system.

Probiotics are not new. They have been used for centuries all over the world. But before the mid-1990s, they went largely undercover in the United States. Many people knew that eating yogurt could help ward off a urinary tract infection, but there was no title for this phenomenon. In 1994 probiotics fell into the spotlight as a new law lowered the bar for premarket safety evaluations of new ingredients used in **dietary supplements**.[1] Suddenly, with less demanding safety requirements, probiotics appeared all over the place prominently advertising health claims. They earned a name and took over entire shelves in drug stores and supermarkets. Today, there are more than three hundred probiotic supplements on the market and hundreds of other products with probiotics in the ingredient list.

Some say that probiotics should be used in tandem with antibiotics to keep up the "good" bacteria supply in the body. Others insist that probiotics are preventive medicines, able to ward off stomach viruses or bouts of diarrhea before they even appear. And a few have made claims that probiotics can—and should—be used by people with chronic diseases. Are any (or all) of these claims true? Could probiotics be panaceas? Are there any dangers associated with using them?

WHAT IS THE DATA?

The human intestine is one of the few organs that put the outside world in direct contact with the inside of the body. Our mouth is a hole— nothing protects our tongue or teeth or gums from the microorganisms that share our earth. It's a short ride down the esophagus, through the stomach, and into the intestines. As a result, the intestinal tract is filled with bacteria, viruses, and other living things, some potentially beneficial and others potentially harmful.

The intestine is lined with a series of protective layers. Minute by minute, the gut has to decide what nutrients should be let in and what toxins or pathogens should be kept out. The normal bacteria in the gut (or microflora, also called the "good bugs") help in this process. The microflora also break down vitamins, fibers, and carbohydrates that were not digested earlier and actually integrate them into the intestinal lining. And specific bacteria produce vitamin K, a chemical critical to proper blood clotting. Without vitamin K, we would not be able to stop bleeding, creating a potential crisis each time we bruised or cut our skin. So beyond playing a key role in gut function, the good bugs also help to physically protect the intestine and maintain health in our bloodstream.

A probiotic is a live microorganism that populates the bowels and benefits the health of the intestine and other organs. In other words, it is a "good bug." Just like microflora, probiotics live inside the human body and improve how our body functions. The only difference between a probiotic and normal microflora is that you can buy a probiotic at the drugstore or supermarket.

To qualify as a probiotic, an organism has to meet several criteria. It must be able to stick to, survive, and multiply in the intestine (this includes surviving the acids in the stomach and intestinal tract); affect the way the immune system responds and functions; influence the biochemical processes of the body (in a good way); and remain viable regardless of how it is given, whether it is a powder added to foods or a yogurt with bioactive cultures. Finally, a probiotic cannot have any pathogenicity: in other words, it is a bug that cannot have the potential to cause illness.

The probiotics sold in the United States include one or more of the following types of bacteria: *Lactobacillus, Bifidobacterium, Enterococcus, Escherichia,* and *Saccharomyces* (the last of which is actually a yeast). *Lactobacillus GG* is the most helpful in resolving diarrhea caused by viruses, especially rotavirus. Bifidobacteria are the type of bacteria found in the intestines of breast-fed babies, and some people believe that their presence accounts for why breast-fed babies are generally "healthier" than formula-fed babies. Probably the best-known probiotic is acidophilus, a type of lactobacillus. But despite its celebrity, acidophilus is not the most effective probiotic.

As evidence has emerged showing that probiotics are probably really good for us, these bugs have started to show up all over the supermarket. You can buy them in the vitamin aisle and in the dairy case too. Milk products like yogurt often claim to have "live active cultures," another way of saying probiotics.

It is great that probiotics seem to be readily available to the average consumer. But just because your yogurt has probiotics doesn't mean you are getting a health benefit. For a probiotic to achieve all of its promised benefit, enough bugs have to get to their destination. It is fairly easy for biologists to figure out which bugs are good and which are bad, but it is not so easy to guarantee delivery of enough good bugs into the intestine. Most researchers think that the minimum effective dose of a probiotic is 1 billion **CFUs** of a certain organism per day. However, there are studies that suggest that 100 billion CFUs per day are necessary to get a health boost.

This is a big difference. And even getting to 1 billion CFUs per day is difficult. (Can you imagine getting your child to eat one billion of anything that is healthy, no matter how tiny?) This is why lots of products on the market that make probiotic claims don't have true probiotic benefits. A yogurt may have live active cultures, but in many cases there are not enough of them to benefit the eater. Essentially, the dose is too low. It would be like giving your child Tylenol or Motrin for a fever but grossly underdosing the medicine—if you gave the right amount, your child would feel better, but with too little the medicine makes no difference at all.

Part of the problem with dosing probiotics is that the amount your child gets depends on the amount he is willing to eat. The other problem is that the FDA does not regulate probiotics the way it regulates medicines. Probiotics are considered dietary supplements, and many have been categorized as GRAS, which means the FDA believes that they have been used long enough or widely enough to have proven their safety without specific study. In several cases, their GRAS status has been backed up by formally collected data suggesting very limited potential for harm. Therefore, these probiotics may be sold in a less-regulated fashion than prescription medicines.

Probiotics still carry potential safety risks, though. Children with compromised immune systems or a history of recent bowel surgery are susceptible to infection from probiotics. One of the defining features of a probiotic is that the bug is nonpathogenic, which means that it cannot cause illness in a person. It turns out that in these two groups of children, certain probiotics seem to have the ability to cause disease.[2] These diseases are quite serious, including infection of the blood (bacteremia) and overwhelming infection of the entire body (sepsis). While this is a very uncommon scenario, parents of children who fall into one of these groups need to be informed. Otherwise, moms and dads may give their child a probiotic, believing that they are doing something good while inadvertently causing harm.

The classification of probiotics as dietary supplements has another downside: their purity and content varies considerably. This is because dietary supplements are subject to a lower level of scrutiny than substances classified as drugs. With a drug, the quality and dose are checked before bottling. Not so with a dietary supplement. The FDA does not analyze the content of dietary supplements; the only checking is by consumer advocacy groups. This means that your child may not be getting what you think she is getting. So if one probiotic doesn't work for your child, it doesn't mean that another won't; just because two labels make similar claims doesn't mean that the bottles contain the same dose of probiotic. Unfortunately, this is the nature of dietary supplements. Regardless, Americans have flocked to probiotics. These products seem to make a difference for people, evidenced by the fact that the industry is growing rapidly. There are two main reasons why people use probiotics. Some are trying to treat acute infectious diarrhea (often called "stomach flu"). As it turns out, if a virus causes the diarrhea, then probiotics tend to be helpful. But if bacteria are the cause, there is no benefit, though there seems to be no harm either. Most people never know what particular infection is causing their diarrhea, and they are willing to try anything—including probiotics—on the chance that it will make a difference.

The second main reason why people use probiotics is that they are trying to deal with the diarrhea that comes with taking antibiotics for an unrelated problem. In this instance, probiotics tend to work very well. Antibiotics are medicines that either kill or stop the growth of bacteria. If your child has an ear infection or bronchitis or a urinary tract infection, an antibiotic may be prescribed to help treat the infection. Here's the problem: antibiotics typically don't wipe out only one type of bacteria. They eradicate a group of bacteria or sometimes several groups. So while your child may feel better because the bacteria causing the ear pain go away, she may also lose important good bacteria in other parts of her body. Typically, antibiotics strip the gut of some (and occasionally most) of its normal microflora. When this happens, taking a probiotic can replenish the good bugs the gut needs in order to function well.

There may be more uses for probiotics on the horizon. There is mounting evidence that probiotics directly affect the immune system, stimulating it to work harder when the body is trying to fight a particular infection. The opposite is also true: probiotics may turn down the immune system when a person is well, theoretically preventing overstimulation of the immune system and even autoimmune diseases (Gill 2004). Major health organizations like **WHO** and FAO (Food and Agriculture Organization) have jumped on this bandwagon, endorsing the role of probiotics in the immune system.

So if probiotics might help and don't hurt your body, it is probably worthwhile to give these products a try. There are so many on the market, though, how do you choose one? If you go with a stand-alone probiotic rather than a food supplemented with probiotic, choose saccharomyces or lactobacillus. *Saccharomyces boulardii* significantly reduces the frequency and duration of acute diarrhea (Szajewska 2007). It is safe and effective for children two years and older. *Lactobacillus reuteri* improves regurgitation and helps infants tolerate feeding better, especially formula-fed premature babies (Indrio 2008). It actually reduces crying in these kids. As more studies evaluate probiotic supplements in children, this list of proven remedies will grow.

The other option is to choose food products, like yogurt, that contain probiotics. However, here the data is mixed. Several studies show that supplemented foods make no difference. For instance, there is no evidence that dairy products with certain strains of bifidobacteria have any beneficial effect. Likewise, most bio-yogurts claiming to improve antibiotic-associated diarrhea do not actually do so in studies. Some yogurts contain "starter cultures" with specific strains of lactobacillus or saccharomyces. The starter cultures are presumably effective, but when the yogurt is pasteurized, many of the live cultures are killed. So while food products are presumably great vehicles for probiotics, manufacturers haven't mastered delivery of these good bugs.

This is not to say that there is a conspiracy of mislabeling in the dairy industry. To the contrary, there is a concerted effort at proper labeling foods. In order for a yogurt to be stamped with the "live and active culture" seal, the refrigerated product must contain one hundred million viable lactic acid bacteria per gram at the time of manufacture. Unfortunately, this number does not tell you whether that was the number of probiotics in the starter culture or whether that's how many bugs are in there on the supermarket shelf. If a product is pasteurized after the bacteria are counted, many of the starter cultures die. Despite an attempt to accurately label foods, the stamp "live and active cultures" doesn't necessarily mean there are enough probiotic bugs to make a difference.

Probiotics have recently moved into the sphere of infant formulas.

So far, the data suggests that certain bifidobacteria are safe and effective in promoting a healthy infant immune system. But most of this research has looked at stand-alone supplements rather than probiotics actually added to formula. Therefore, these premixed probiotics are probably still safe but not necessarily effective. Once a can of powder formula is opened, moisture gets in and probiotic bacteria begin to die. Even if enough probiotics survive and make it into an infant's intestine, there is no evidence that they are beneficial or at all long lasting (Verkler 2008).

The purpose of putting probiotics into infant formulas is to mimic what happens in the guts of breast-fed babies, whose intestines are colonized with bifidobacteria. Because breast-fed babies tend to have fewer or less-severe illnesses compared with formula-fed babies, some speculate that bifidobacteria hold the key to infant health. Formula manufacturers are betting on this and adding the good bugs to their products.

There are probiotics that can help when you are sick and possibly probiotics that work preventively when you are well. Now there is a whole new class of dietary supplements called "prebiotics." These aren't actually bugs like probiotics but rather nondigestible nutrients (such as oligosaccharides) that stimulate the growth of specific bacteria in the gut. Basically, they are probiotic food. It turns out that breast milk has a high content of prebiotics. So the most current thinking is that probiotics are beneficial, and when prebiotics are added to the mix, the probiotic organisms live longer and do better.

WHAT IS THE BOTTOM LINE?

Probiotics are generally safe. The main exception to this rule is among people who have compromised immune systems or have recently undergone bowel surgery. For all other kids (and us parents), the data so far looks good.

The issue with probiotics is simply that it is difficult to know whether you are taking enough to make a difference. In order for probiotics to be helpful, enough organisms must be ingested. A person needs to consume somewhere between 1 billion and 100 billion CFUs per day. Because probiotics are classified as dietary supplements, there is little regulation over precisely how much live product makes it into any given powder, tablet, or food. So while you might think you are giving your child something beneficial, you may actually wind up underdosing the probiotic and providing no benefit at all.

The effectiveness of probiotics also depends on what you are trying to accomplish. If your child has diarrhea caused by a virus, probiotics will likely be helpful. But if bacteria cause your child's diarrhea, probiotics have no clear benefit. If the diarrhea is a result of antibiotic use, probiotics may work very well to turn things around. As it stands right now, there is no good data suggesting that probiotics can be used on a daily basis to prevent intestinal infections.

Ultimately, whether you get your probiotics at a market, health food store, or the pharmacy, no one can vouch for any given product's purity or efficacy. This is because they are classified as "dietary supplements" rather than as drugs, and therefore are not subject to certain rigors of FDA testing. Furthermore, when it comes to dairy products that have been heat treated or pasteurized, the label may say that it contains good bugs but you don't know if they are live or dead—so there's no way to tell whether the probiotics will work.

Ultimately, probiotics are almost always safe. But because they are considered supplements rather than drugs, no one is required to guarantee the dose, the delivery, or the effectiveness. There should be a move to consider these supplements medicines so they become subject to more stringent standards and can help more people.

WHAT'S IN MY HOUSE?

I use probiotics once or twice a year when my family is fighting off a stomach virus. While they may not help, they certainly don't hurt. The hard part is that with diarrhea most doctors caution to stay away from dairy—which is precisely the vehicle many probiotics come in. I have found some very kid-friendly, odorless, colorless, tasteless powdered probiotics, but I never remember them until my kids are in the throes of vomiting and diarrhea. Honestly, even I have no idea if they work.

Chapter 6

Soy

WHAT IS THE QUESTION?

The controversy over soy has been brewing for decades, though it has only recently secured a fixed spot on the list of things that concern parents most. The fears largely have to do with chemicals called **phytoestrogens** that are a normal component of soy. As their name implies, these chemicals resemble the female sex hormone estrogen. Many people fear that because phytoestrogen sounds like estrogen, it behaves like estrogen in the human body. It's not uncommon for a parent to ask me: if I give my son soy products, will he develop breasts?

The hormonal concerns over soy are largely overplayed, but, ironically, there is a true controversy involving soy that deserves attention. Soy is one of the most common **genetically modified foods** in our diet. The risks of genetic engineering of foods—from the immediate (like allergic reaction) to the longer term (like further increasing rates of antibiotic resistance)—have not hit the mainstream quite yet.

Therefore, when parents question the safety of soy, they often fear the wrong thing. Is soy dangerous? Possibly, but not because of estrogen.

WHAT IS THE DATA?

Soybeans are the seeds of the soy plant. These beans have more protein than almost any other vegetable, and they are the basis of a number of different food products, including tofu, soy sauce, and soybean oils. Soy contains a naturally occurring chemical called phytoestrogen.[1] This chemical looks a lot like the female sex steroid estrogen, hence its name.

Estrogen is a hormone produced by the human body. Both men and women make estrogen, though women make much more of it. There are dozens of hormones naturally produced by the human body, including estrogen, testosterone, thyroid hormone, growth hormone, insulin, dopamine, and epinephrine. Many of them are steroid hormones. These infamously named steroids are just a group of chemicals that share a similar general structure. The steroids that the public tends to worry about are not the naturally occurring ones but rather the manufactured drugs (like prednisone or anabolic steroids). These are called steroids because they have the same chemical backbone as the body's native steroid hormones; yet each member of this club works differently. Some of the manufactured versions have given the name steroid a very negative connotation, but there are really only a few bad actors.

While phytoestrogens are chemical look-alikes to estrogens, and while their name sounds suspiciously similar, phytoestrogens do not behave like naturally occurring hormones in the body. This is because phytoestrogens are *not* steroids. Therefore, they do not affect organs and tissues in the body (like breast tissue) the same way estrogen does.[2]

We know that phytoestrogens are not the same as estrogens. What we don't know is whether phytoestrogens are beneficial or dangerous. Some studies have shown that soy is protective against heart disease, thyroid cancer, and even breast cancer, especially when a person is exposed during childhood (Horn-Ross 2002; Korde 2009). Other studies conclude the opposite: soy's phytoestrogens pose dangers to reproductive function, immune function, and the thyroid gland, and they can even accelerate the growth of some tumors (Gallo 1999; Yellayi 2002; Luijten 2004). It is important to keep in mind, however, that most of the studies claiming harmful effects were done on lab animals rather than on humans.

Children get soy in food and drink. The exposure can start very young, in the form of infant formula.[3] Infants who drink soy formula have been found to have high levels of circulating phytoestrogens. These levels can be shockingly high—up to ten times greater than the levels in adult women eating high-soy diets (Setchell 1997). For good reason, some parents who heard this panicked.

In 1998, the **AAP** released a statement that even if phytoestrogens have effects on rodents and other lab animals, they don't work that way in humans. The phytoestrogens from soy don't attach themselves very well to human estrogen receptors, so they aren't very potent in human bodies. The AAP tried to reassure parents that phytoestrogen levels of infants drinking soy formula may be high, but the chemical effect on their bodies is negligible.

But more than ten years after the AAP statement was published, there is still controversy surrounding the hormonal effects of soy. The difference today is that soy is no longer being questioned in isolation. It's not just about the phytoestrogens anymore. Instead, a larger group— called endocrine disrupters (or **EDs**)—has replaced phytoestrogens on the list of most feared chemicals.

EDs include phytoestrogens as well as others like **phthalates** in cosmetics and **bisphenol A** in plastic containers. These are chemicals that may affect the way hormones work in some people's bodies. EDs are not hormones themselves but rather are chemicals that look an awful lot like hormones, mimic them, or affect one little step in the complicated cascade of hormones in our bodies. As a group, EDs have been implicated in all sorts of biological heists, from the early onset of puberty to cancer formation. Phytoestrogen itself has not been clearly linked with hormonal effects, but its umbrella organization of EDs has.

So in 2008, when the AAP published a follow-up report on soy formula safety, this one was much more detailed on the issue of phytoestrogens. The conclusions were the same: you don't need to worry about phytoestrogens if you are giving your baby soy. But the report was silent on the larger subject of EDs.

I personally don't worry about soy formula and phytoestrogens. I think the data is pretty reassuring, and I've watched lots of kids grow and thrive while drinking soy. Where I think matters get complicated is when we turn away from hormones and chemicals innate to soy and instead focus on farming and manufacturing practices. Genetic modification of various foods—like soy—has fundamentally changed the nature of these plants. I really believe that the soy of the past, including its phytoestrogens, has no clearly proven harmful effects. But the soy of today is a different genetic beast.

A description of how foods are genetically modified also appears in the chapter about allergies, but I think it bears repeating because it affects a huge portion of our food supply, and because the process is pretty difficult to grasp on the first read-through.

Over the past two decades, agricultural scientists have genetically modified the DNA of a variety of crops, including soy, corn, cottonseed, canola, tomatoes, potatoes, squash, carrots, sugar beets, papaya, wheat, rice, and milk and dairy products, among others. The genetically modified products have been available for purchase in the United States since 1996. The process of genetic modification involves inserting new genes into the DNA (also called the genome) of a particular crop. The motivation behind genetic modification of foods included increasing bounty, decreasing pesticide use, and enhancing the nutritional benefit of crops around the world, all praiseworthy goals. But if genetically modified foods have new parts to their DNA, their effect on the human body may change. Genetically modified foods may be making us sick.

The mechanics of genetic modification are complex. When an agricultural scientist wants to modify the genome of a plant, new DNA must be artificially inserted into the plant's genetic code. First, the scientist needs to design a specific gene, for instance a gene that allows a soybean to withstand pesticide sprays. This is the easy part.

Next, another gene (called a "promoter gene") must be attached to the first gene in order to achieve insertion. A new gene cannot just integrate itself into DNA; it needs help. The promoter gene is like scissors and tape—it can splice itself and anything attached to it into the genome.

Finally, a third gene must "mark" the new DNA. This helps scientists identify which plants were successfully modified. A marker gene might be one that resists antibiotics, so when scientists want to see whether their genetic modification worked, they can sprinkle seeds with antibiotics—the ones that survive presumably have the marker (antibiotic-resistant) gene and the promoter and the gene that they wanted in there in the first place. So when we talk about genetically modified plants, we aren't just talking about adding one gene into the mix. Rather, we are talking about three or four added, sometimes more.[4]

Here's the dirty little secret: genetic modification isn't uncommon. Today 92 percent of all soybeans in the United States are genetically modified—and soy isn't alone. Not only is there a long list of genetically altered foods, but when you look at a particular food on that list it can be difficult to find a non-genetically altered version: 80 percent of U.S. corn, for example, and 86 percent of cotton are genetically modified.

So what if soy is infused with new genes? Does it really affect us? The answer is that we don't know, but there is a strong possibility it does. Genetic modification of foods seems to be associated with the increasing prevalence of food allergies. Many speculate that this is because the promoter gene is made from a different food type than the food into which it is being inserted. For instance, a peanut gene may be used in the genetic modification of a squash. Suddenly, a person who is peanut allergic may be at risk when eating squash.

Even if the peanut promoter gene does not trigger an allergic reaction (and in many cases it won't), there are people who are very diligent about avoiding certain foods. These people may be inadvertently exposed to these foods—or tiny pieces of them, to be more specific—by eating a genetically modified food containing part of their genome. In other words, a peanut-allergic person eating squash isn't thinking about peanuts.

Another issue related to genetic modification concerns antibiotics. As described, the marker gene identifies whether a new DNA sequence was adequately incorporated into the genome. Often marker genes are made from antibiotic-resistant genes. This is an easy way for scientists to test whether their modification worked. The only problem is that genetically modified plants have also now incorporated antibiotic resistance. If the gene in the food stays in the human body after the food is consumed, and if the gene remains active (both big ifs but theoretically possible), it can potentially make the human consumer resistant to that specific antibiotic. On a larger scale, this can increase the prevalence of antibiotic resistance in our population.

The final issue with genetic modification isn't personal—it's global. Over the past two decades, as genetic modification of foods has been incorporated into American farming, countries around the world have decreased their imports of American agricultural products. In the European Union, for example, genetically modified foods are not welcome.[5] Beyond the potential health implications, therefore, are economic repercussions of this burgeoning agricultural technique. And while others are turning away our genetically modified exports, we consume more and more of them. To put the American consumption of genetically modified foods into perspective, consider this: as of 2003, 90 percent of the money Americans spent on food was for processed products, and 70 percent of those processed foods contained ingredients that had been genetically modified. There is currently no law in the United States mandating that genetically engineered foods be labeled as such, so most Americans remain completely unaware of how many genetically modified products they are ingesting.

The dilemmas that arise from genetic engineering of our foods are major. They cut across medical and moral lines, and verge on science fiction. In the words of Doris Rapp, author of *Our Toxic World*: "Without labels, those who have known food or other allergies, those whose religion includes food restrictions, those in favor of animal rights and those who are vegetarian have a right to complain. Does the vegetarian want to eat tomato that has been combined with fish? Does the peanut-sensitive individual want to eat hidden peanuts in some food product? Does the Orthodox kosher Jew want to ingest milk in some altered meat product? Does a Hindu want to eat beef or milk in some hidden form? Does the Christian Scientist want an antibiotic marker in his intestines? There is no way of determining what is acceptable for them to ingest. They have lost a basic right of freedom of choice."

Ultimately, we do not know how genetically modified foods impact our bodies. It is likely that they contribute to allergies; perhaps they will increase antibiotic resistance. But because these are uncharted waters, the long-term effects are unstudied. It is almost impossible to assess them since each modified food is crafted in its own particular way. Consumers always want to know if something is a carcinogen—if it can cause cancer. Even though there is no current data showing that genetically modified soy is dangerous in this regard, we don't know if it is safe either.

WHAT IS THE BOTTOM LINE?

Organic soy produce has never been proven to have estrogen-like effects. There are entire countries whose populations subsist on soy as their main source of protein, and the men are not running around with breasts.

Therefore, from an estrogen standpoint, soy is safe. There are some emerging concerns about soy's effect as endocrine disruptors, or chemicals that affect the way hormones work in some bodies. But specific endocrine-disrupting effects of soy have not been clearly linked to serious diseases.

A much bigger issue, in my opinion, is that of genetic modification. As of the writing of this book, 92 percent of the soy crop produced in the United States contains genetically modified soy. We don't know all the risks, but they include allergic reaction, exposure to genes that promote antibiotic resistance, and even ethical dilemmas, because soy often now contains genes from other food sources.

There is no labeling requirement for genetically modified foods. But we are all exposed, one way or another, save those who grow all their own produce, drink milk from their own grass-fed cows, and live on a self-sustaining farm. The rest of us cannot avoid exposure. Given that we do not know the implications of genetic modification, those who want to be able to avoid it should be given the chance. For people who want to make a change here, this means fighting for labeling on all foods.

In the meantime, buying "**certified organic**" (also called "**non-GMO**," for non-genetically modified organisms) foods will minimize your exposure to genetically modified soy. It doesn't guarantee that there is no altered DNA in the food itself, because neighboring farms often cross-pollinate— if one farmer is certified organic but his neighbor plants genetically modified crops, then wind, insects, and other forces of nature will mix the genetic pools. But it still significantly reduces your risk of exposure to modified DNA when sipping on a soy latte or snacking on edamame.

WHAT ARE EDAMAME?

In Japanese, *edamame* means "beans on branches." A more useful translation is "boiled soybeans." Once confined to Japanese restaurants, edamame are now fairly commonplace in supermarkets across the United States. Sometimes they come inside the pod (so they look like short, squat string beans), while other times they are served out of the pod as stand-alone beans. These beans are rich in fiber, protein, omega-3 fatty acids, calcium, and vitamins A and B. For kids who aren't big on green vegetables, this is one worth trying because edamame don't look, taste, or smell like most other green veggies.

WHAT'S IN MY HOUSE?

Soy products can be miracle cures for the intestine. They can stop diarrhea and fix constipation too. My daughter was a very constipated toddler but did much better when I switched her from cow's milk products to soy. I never worried about the phytoestrogens, but I did buy organic products. These days, my daughter's back on dairy. But my freezer is stocked with edamame, we are occasional soy milk drinkers and tofu eaters, and I never sweat it. I just choose the organic variety when I shop.

Chapter 7

Vitamins and Supplements

WHAT IS THE QUESTION?

The question is pretty straightforward: does my kid need to take a vitamin? Vitamins and nutritional supplements are appealing because they are "natural." Adults use them often in an attempt to prevent illness when they are well and to cure it when sick. Vitamins and supplements seem pretty harmless because they aren't "medicines." Many have rays of sunshine or happy faces on the labels—so how dangerous can they be?

Even among physicians, there aren't many concerns about dangers of vitamins. It is possible to overdose or to have a serious reaction, but this is pretty rare. In fact, with vitamins and supplements the issue gets turned on its head: if these products are almost completely harmless, do they even work? Most vitamins and supplements are considered safe, but are they effective?

For our kids, the lure of vitamins goes beyond their presumptive healing properties. A picky eater may not be getting the vitamins and minerals she needs through her diet. There are kids who literally eat only yellow or white foods (pasta, cheese, bread, banana, maybe some chicken nuggets). For these kids, vitamins would seem important. But do they make a difference? Do they fill the gaps in your children's diet and help round out their nutrition? Or do the supplements just go straight through them? And with hundreds of varieties on the shelves, how do you choose the right one?

WHAT IS THE DATA?

We all know that the twenty-first-century diet is less than ideal. In the old days, before industrialization and processing and preservatives, the human diet was much higher in nutrients and lower in calories. Fresh fruits and vegetables were staples. Humans also exercised more; just getting around and tracking down a meal required exercise. Sure, our ancestors lacked access to modern medicine, and their average life span was decades shorter than ours now. But still, generations before us did some things better than we do today.

Vitamins and supplements try to make up for some of our modern-day nutritional deficiencies. You can buy just about every nutrient imaginable: vitamins A, B, C, D, E, and on through the alphabet; commonly known minerals like calcium, magnesium, and phosphorus; obscure ones like colloidal mineral complex; amino acids (arginine, carnitine, and glutamine, to name a few); fatty acids like cod-liver oil, flaxseed oil, **DHA**, and omega-3s; even supplements targeted at specific symptoms and sold under names like Blood Builder, MegaFood MegaZymes, and Liver Support. There are thousands available.

For people who don't eat a balanced diet, vitamins and supplements offer to fill the nutritional hole. You can buy them just about anywhere, and they cater to people of all shapes and sizes. For the health-conscious adult, there are impossibly large horse pills that smell just a little bit like freshly mowed grass; for infants there are liquid drops; for the person with a sweet tooth there are caramel chocolate chews. Kids may not get their fair share of fruits and vegetables, but our sons and daughters can get what's missing in one tasty chewable tablet shaped like their favorite cartoon character.

Just in case you don't want to buy your vitamins, they are also integrated into almost all the processed (and even nonprocessed) foods you eat. They are added to cereal, bread, rice, pasta, milk, yogurt, cheese, orange juice, and the list goes on.

But don't be fooled: supplements may add vitamins or minerals that you don't eat enough of, but these products aren't an alternative to a healthy diet, and they don't rationalize eating junk food. You can't eat an unhealthy diet and then balance it out by taking vitamins.

Almost a third of all American kids take supplements, mostly in the form of multivitamins. This may sound like a lot, but in the 1970s about half of all kids took them. It is really the parents who push vitamins; many of us buy them and encourage our kids to take one every day. And it is widely accepted that vitamins are good for kids. Even parents who don't buy them tend to believe this is true.

So are they good for you? Sort of. It looks like vitamins aren't *bad* for you. Studies of multivitamins show that there is no real harm, and it is very rare for a person to have a reaction. But just because vitamins aren't bad for you doesn't mean that they are *good* for you either.

The first reason there is no clear benefit to taking multivitamins is that these products may not do what the label promises or contain the ingredients precisely as listed. Vitamins are not medicines; therefore they are not regulated as such. They fall under the purview of the Dietary Supplement Health and Education Act (**DSHEA**) of 1994, which eliminated the FDA's premarket review of dietary supplements. Under the DSHEA, manufacturers could make unproven claims. Now, this isn't to say that false advertising is allowed—manufacturers are legally bound to list the ingredients on the label and are held accountable if it is discovered that the product doesn't contain what the label says it contains. They can also be censured if health benefits are promised and turn out to be completely misleading. But that's pretty much the extent to which manufacturers of supplements have to comply: if they are caught, the companies are in trouble. On the other extreme, every medicine on the market is tested prior to FDA approval. Medicines are subject to random and frequent testing in the lab before they are even put into the bottle. When it comes to medicines (as opposed to vitamins), the quality assurance standards are much higher and the health benefit claims much better vetted.

Therefore, when a medicine is found to have the wrong dose in a capsule—which happens about a couple of times a year—it is quickly recalled. But in order to figure out whether a vitamin or supplement has the wrong dose in a capsule, a consumer group has to proactively go out and do the testing.

This testing has been done many times over, and each time the results look similar. Surprise, surprise, often the dose of vitamins or supplements is not as advertised. One recent study performed by the pharmaceutical watchdog ConsumerLab found that among multivitamins sold in the United States and Canada, fewer than half actually contained what the labels said they did. Most of the time, the dose is well below what the label says. But ConsumerLab has documented many cases erring on the other extreme, like one multivitamin for children that contained more than twice as much vitamin A as indicated on the label.

I always assume that if a vitamin or supplement is mislabeled it probably contains too little of an ingredient. Somehow it never crosses my mind that a manufacturer would put *more* of an ingredient into a formulation. But it clearly happens both ways. And an excessive amount can be dangerous. If your child's multivitamin has too much vitamin A, for instance, the excess can ultimately cause weakening of the bones and even problems in the liver.

The risk of overdosing vitamins or supplements is more theoretical than practical. There are very few case reports of children who took a multivitamin as prescribed on the label—usually one a day—and wound up with a severe nutrient overload. Teens and adults are the ones at greater risk, because they often choose to take more than directed, believing it is impossible to have too much of a good thing.

By far, the more common scenario is underdosing. The label promises that a multivitamin has certain doses of specific ingredients when in fact it may have only a fraction of the advertised amount or none at all. If you have ever begged your child to take a vitamin and fought that good fight, the notion that you are giving your child a watered-down version of what you think you are giving them will frustrate you to no end.

Ultimately, given the labeling standards for vitamins and nutritional supplements, you will never know for sure if what you think is in your vitamin is really in your vitamin. It helps to have a seal of approval from the United States Pharmacopoeia (USP), NSF International, or ConsumerLab: this means that one of these organizations actually checked (at least a representative sample of) the vitamin and agrees that what's on the label is actually in the bottle.

The second reason there is no clear benefit to taking multivitamins is that the human body may not absorb them efficiently. Sometimes— maybe even often—vitamins go in one end and come out the other. So let's say you take that leap of faith and decide to believe that when you buy a multivitamin for your child you are giving the vitamins as advertised. (I mean, at some point you have to have faith that manufacturers of many of these products are really trying to help people and not just out to make a buck.) What happens to these vitamins and supplements in our bodies?

Chemists can easily isolate or build specific compounds like vitamins, fatty acids, and minerals. But it can be very tricky to formulate them in a way that is absorbable by the body. Even if nutrients can be packaged so the intestinal tract can effectively extract them, all our other medicines, supplements, and even foods can interfere. People who take supplemental vitamins often assume that the amount their body extracts from a given pill and the amount their neighbor's body extracts from the same pill is equal. This is not true.

It turns out that the absorption of any vitamin depends on a variety of factors: age, gender, the amount ingested, the chemical formulation (including whether it has fillers or coatings), the mix of vitamins, whether you are taking it on an empty or a full stomach, and any other medicines you are currently taking.

Now, vitamins and minerals within foods—the vitamin C inherent to oranges or the omega-3s that are naturally a part of salmon—are certainly not absorbed consistently either. Your neighbor may get more calcium from his black beans than you get from yours. But, in general, the delivery of vitamins and minerals is better when *included* as part of a food. For this reason, many foods on the market are now fortified with vitamins or minerals. Orange juice often has added calcium and vitamin D; eggs come with added omega-3s.[1]

Vitamin manufacturers use the same rationale to enhance the absorption of their supplements. Most multivitamin labels suggest that you take the vitamin with a meal. If you are not eating the nutrient as actual food, you might as well try to trick your body into thinking you are.

But this still doesn't mean that the vitamin will be absorbed. For instance, we know that calcium is critical for our bones and teeth. If your child refuses to eat calcium-rich foods (dairy products, green veggies, black beans, and orange fruits among them) and won't drink milk, it is only logical—or, at the very least, hopeful—to assume that you can make up the difference by giving her a calcium supplement. But there can be a big difference between the absorption of calcium in milk and the absorption of calcium in a multivitamin. If your child's multivitamin has both iron and calcium, the iron will actually block the absorption of calcium, undermining the entire purpose of your supplement.[2] Vitamin B_{12} has a similar story. If a multivitamin has more than 500 mg of vitamin C and it has vitamin B_{12}, the vitamin C at this dose will block the absorption of the vitamin B_{12} (less than 500 mg of C won't). Competing micronutrients aren't rare: you might think that a multivitamin is giving your child twelve or fifteen vitamins when in fact only ten are getting in.

Then there are the macronutrient stories: the claims that vitamins literally pass straight through us. There are tales of hospital nurses who have nicknamed multivitamins "bedpan bullets" because they find them intact in the bedpan after passing through the patient's entire intestinal tract. There are reports of multivitamins found with the brand name still visible on the pill in the collection tanks of portable toilets from construction sites, apparently after they have passed straight through someone's gut. These anecdotes are not reported in scientific journals, but stories abound. If these are true, well, then it's just an example, albeit at a more dramatic level, of malabsorption.

Despite the reality that some—and maybe even *all*—of a vitamin goes straight through the intestine, I still believe there are groups of children that should take them. Kids in developing countries are the best example. Impoverished underdeveloped nations are filled with children who face dramatic nutrient deficiencies, and multivitamins have been shown to make a big difference for those kids. Even though only a small amount of vitamin is consistently absorbed, some is much better than none. In the United States, children with chronic illnesses are often prescribed vitamins or supplements specific to their diseases, and these improve their nutritional status.

How about a typical, healthy kid here in the United States? Does that child need to take vitamins? You might think I would say no because it is unclear whether vitamins make any difference at all. But surprisingly, the answer is probably yes, particularly if you have an infant, a teenager, a self-proclaimed vegetarian, or a superpicky eater. Here's why.

I'll start with the case for infants. In 2008 the **AAP** recommended a multivitamin specifically including vitamin D for all breast-feeding infants and for formula-feeding infants drinking fewer than thirty-two ounces of formula per day. This recommendation was based on a need for more vitamin D—400 IU per day, to be precise—in order to keep bones strong and to prevent rickets, osteoporosis, and fractures.

Sunlight is actually our primary source of vitamin D, but the AAP recommends keeping babies out of direct sunlight and using sunscreen to protect them when they are outside. This recommendation comes from the belief that risk of later skin cancer is greater than the vitamin D benefit from the sun, especially if you can get that vitamin D from elsewhere. In order to get enough vitamin D from sunlight, your skin must be fully exposed. Skin protected with sunscreen cannot produce as much (some would say any) vitamin D. Even if you decide to ignore this recommendation and take your baby into direct sunlight, if you live in certain parts of the country—particularly in the northern United States—the **UV** radiation in the winter is insufficient to produce enough vitamin D. Most babies start cow's milk when they are a year old. Cow's milk is fortified with vitamin D, obviating the need for multivitamins. But before this, if your child gets the majority of her milk calories from breast milk, she probably isn't getting enough vitamin D. By the way, the need for vitamin D doesn't go away, and fortified milk is actually the primary dietary source of the vitamin for people of all ages, adults included.[3]

So if infants need vitamin D, why recommend a multivitamin? Because, at least for now, vitamin D can be hard to find on its own in an infant-friendly dropper formulation. Therefore, you are stuck with a multivitamin. This is fine—again, there are no real safety issues with the other vitamins mixed into the infant drops—but if you have ever given your baby multivitamin drops, it won't be news to you when I say that these vitamins don't taste great, especially the ones that contain vitamin B.[4]

That's the story for infants. What about older kids? Well, if you have an outrageously picky child (like the one who only eats yellow foods), a teenager, or a vegetarian in your midst, the answer is that a multivitamin won't hurt. It may not help much either, but if you can round out the nutritional balance for your rapidly growing child and there is a chance that there will be a benefit, why not do it?

Most multivitamins designed for children and teens advertise the recommended daily allowance (RDA) of thiamin, riboflavin, niacin, folic acid, and vitamins A, B$_6$, B$_{12}$, C, D, E, and K. If your child is a vegetarian, supplementing vitamin B$_{12}$ is especially important. We get it mostly from meats, so a child who doesn't eat meat really needs to make up the difference with fortified drinks, cereals, or supplemental vitamins. (Vegetarians also run the risk of consuming too little protein. If they choose not to eat meat or fish, they can meet the RDA with dairy products and legumes.)[5] I generally think of self-proclaimed vegetarians as the ones most in need of vitamins because vegetarian families almost always know more about basic nutrition. Parents who are committed to eat nonmeat proteins have generally read about how to balance their diet. Some vegetarian parents have seen nutritionists. All will pass this information along to their kids. More often it is the lone vegetarian within a family who stands to become deficient in certain vitamins and minerals.

As for teenagers in general, they almost always need more calcium than they get when left to their own devices. Teenage girls also have an increased iron requirement because with their periods begins a loss of blood, and without supplemental iron, girls can slowly become anemic. The biggest issue for teens is that their lifestyle tends to compete with balanced nutrition: these kids wake up late and often skip breakfast, buy lunches typically loaded in carbs or high in sugar (or both), and prefer the accessibility and the price point of fast foods and sodas. The ideal balanced family dinner doesn't necessarily happen if our teens are practicing with a sports team, engaged in extracurricular activities, holding an after-school job, or simply trying to get through their nightly homework load.

If you think about it, the dietary habits of a superpicky toddler are not that different from those of an adolescent: carbohydrate intensive with a smattering of sugar. Getting fresh fruits and veggies or protein into a member of either of these groups can be a battle. For nearly identical reasons—arrived at very differently—picky toddlers and teenagers are both good multivitamin candidates.

The possible benefit of multivitamins doesn't end with adolescence: these vitamins are probably good for us adults too. In our version, though, folic acid takes center stage as one of the most important ingredients. Folic acid became famous because of its role in prenatal vitamins: it reduces the risk of neural-tube defects in fetuses by 70 percent. This particular vitamin likely has other benefits too: if you consume a high level of folic acid daily, you may lower your risk of colon cancer or breast cancer.

But overall, whether we are talking about toddlers or teenagers or adults, vitamins aren't crutches. Remember, their absorption is very variable. Food choices, exercise, and general lifestyle habits make a far bigger difference to our health and well-being than taking a vitamin. Vitamins don't compensate for the massive risks associated with smoking, obesity, or inactivity. Because multivitamins are cheap—it costs about as much for a multivitamin as it does for a quarter of a serving of fruits or vegetables—there is little downside to adding them. But don't fool yourself: these supplements are not going to save your life.

So at the end of the day, there is no real evidence that multivitamins help, but there is no evidence that these products hurt either. Some parents will decide not to waste the money or to pick a more worthwhile battle. Others choose to include multivitamins in the daily routine since there is no harm and some possibility of benefit. Both options are fine and not the least bit unreasonable. But if you are one of the parents who are going to try to give regular multivitamins, how do you choose among the dozens of brands out there?

I think this is much more an issue of practicality than of branding. First of all, I know of no studies comparing all the different brands of children's multivitamins, so endorsing one is impossible. Second, if you can't get the vitamin in, the game is over. So choose a vitamin based on what your child will be willing to take.

If you have an infant, you will learn quickly that babies spit things back. I don't think I fully appreciated this when I first started practicing pediatrics—in fact I cringe when I think about all of the times I told parents, "Just get a medicine dropper and squirt it in his mouth; it's no big deal." Little did I know that babies have no qualms about pelting whatever you give them right back at you. (Thanks to my son, I learned that lesson many times over.) Most infant vitamin drops taste pretty bad. So you may have to hide the vitamin solution in milk or, once your baby is old enough, in semisolid food.

For toddlers and kids, chewable vitamins are a much easier sell. There are two varieties: solid ones that can be chalky or gummy vitamins that can stick to the teeth. Dentists strongly prefer the chewable variety, but most three- and four-year-olds beg for the gummy vitamins, which, incidentally, seem like the lesser of the two choking hazards.

The sooner you can teach your tween or teen to swallow a pill, the better—because these kids can take a tasteless pill form. There are candylike vitamins available, particularly in the calcium supplement department. I worry about these a little because some teenage girls have told me that these supplements taste so good they will eat a few every day. The girls think there is no harm, but overdosing on calcium can cause problems like kidney stones.

WHAT IS THE BOTTOM LINE?

Multivitamins are safe. For the average healthy person, they aren't going to save your life, but they won't cause problems either. If there is even a chance that a few of the components of a multivitamin are going to make their way into the body and strengthen the bones or protect the heart, why not add them to your daily routine?

But just because your kid is willing to eat a multivitamin every morning, don't give up trying to encourage him to eat healthier foods. Multivitamins are not substitutes for healthy living. A balanced diet and exercise will go a lot further than a daily supplement.

WHAT'S IN MY HOUSE?

My kids don't take vitamins. They eat pretty balanced diets and I see no need. But if I happened to have one of those superpicky kids, I would give her a daily multivitamin. I took prenatal vitamins for several years. At the time, I credited them for my rapidly growing hair and strong nails. I felt so great that I tried to convince my husband to take them. (You can probably imagine his response to urgings that he take a prenatal vitamin.) In hindsight, I think the euphoria and hormone shift of pregnancy also played fairly large roles: it wasn't only a vitamin-induced effect.

Part II

Drinks

Chapter 8

Caffeine

WHAT IS THE QUESTION?

We are all intimately acquainted with caffeine. There are some of us who adore it, others who avoid it like the plague, and people in between.

Caffeine is a well-known stimulant sold for three or four dollars a cup at Starbucks and other coffeehouses around the country. It is also in many types of teas, sodas, energy drinks, and even certain foods, namely, chocolate-based sweets.

Adults feel a certain amount of control over their caffeine intake, but we often overlook it when it comes to our children. What is caffeine doing to kids and teenagers? Does it really stunt their growth? Will it truly keep them up at night? We know it's not great for us, but is caffeine actually dangerous for our kids?

WHAT IS THE DATA?

Caffeine is a natural stimulant found in certain plants like the tea bush, the coffee plant (the coffee bean is the pit inside its fruit), the cacao tree (whose seed produces cocoa), and the kola tree of the African rain forest (with its kola nut). Other less well-known caffeine sources—at least, less well known to Americans—are yerba maté and guarana berries. Caffeine is a member of the larger family of natural stimulants called methlyxanthines (sometimes just called xanthines). The **FDA** requires caffeine to be listed on a label when it is *added* to a food but not when it is a naturally occurring ingredient. That's why caffeine isn't listed on the label of a chocolate bar but it is on the back of a soda.

In nature caffeine serves a very important purpose: it protects the plants that manufacture it by acting as a pesticide. When animals nibble on coffee plants or tea leaves, the caffeine can interfere with their ability to feed effectively and in many cases can paralyze or even kill bugs. So caffeine is a natural defense system designed by plants to ward off predators.

Humans, of course, are not at risk of death by caffeine consumption. Or, more precisely, it is difficult for a human to eat or drink so much caffeine for it to be acutely toxic.[1] It would take somewhere between eighty and one hundred cups of coffee, consumed over a relatively short period of time, to kill you.

For us (as opposed to insects), caffeine serves a very helpful purpose: it is a stimulant. Caffeine directly affects the neurons in our brain, making us feel more energetic and more awake. It works by competing with adenosine, a chemical normally produced in our body. Chemicals that affect the way nerves function are called neurotransmitters. Adenosine is a neurotransmitter with a vague job: it slows us down. It can lower heart rate, make us sleepy, and even turn down the immune system by acting as an anti-inflammatory. Adenosine has effects all over the body, not just inside the brain.

Caffeine is similar enough to adenosine in size and shape that it competes in the body with it. On the outside of every cell are receptors that fit specific chemicals, such as hormones, neurotransmitters, or drugs. Caffeine fits neatly into the adenosine receptor, filling it but not telling the cell to turn down. Meanwhile, there is adenosine floating around the body, but without available receptors, the adenosine cannot do its intended job. So by occupying adenosine receptors, caffeine essentially has the opposite effect of adenosine—it turns the cells "up" instead of "down."

Caffeine works best at low to moderate doses (50–300 mg), increasing alertness, energy, and concentration; speeding and clarifying the flow of thought; increasing focus; and improving coordination. There is also research suggesting that caffeine can prevent certain illnesses. One series of studies shows that caffeine consumption may protect against the development of Parkinson's disease (Chen 2001; de Lau 2006).

But caffeine is not a panacea. Anyone who drinks coffee knows there is such a thing as drinking too much. At high doses, negative effects begin to take over: excessive caffeine can make you jittery or nervous; you can have trouble sleeping; it can make you feel anxious or restless; it can cause problems with concentration; it can upset the stomach; it can increase your heart rate (sometimes making it feel like your heart is pounding) or raise your blood pressure; and it can cause headache. This is generally not enough caffeine to do serious damage, but it will be enough to feel awful.

It turns out that not everyone has the same sensitivity to caffeine. Have you ever been out to dinner and when it's time to order coffee someone says that they absolutely must have decaf because caffeinated coffee will keep them up all night? It is possible that this is not an exaggeration. Some people are exquisitely caffeine sensitive.

Even at low doses, caffeine has negative effects. It functions as a diuretic, meaning that it makes us urinate more. This can cause dehydration. The dehydration from caffeine is usually mild, but sometimes it can make a person feel pretty rotten. People who get migraine headaches often complain that caffeine is a trigger. Ironically, it can also help treat migraines, so many over-the-counter headache medicines actually contain caffeine.

There is also research that links caffeine with broader health problems. A few recent studies suggest that even at relatively low doses caffeine increases a woman's risk of miscarriage (Weng 2008), though this claim has been offset by studies concluding the exact opposite (Savitz 2008).

When it comes to bone health, no one seems to agree on the role played by caffeine. It has long been blamed for osteoporosis (bone loss) in women. The association between caffeine and osteoporosis initially came from two independent pieces of data: the first showed that the more caffeine a woman drinks, the more calcium she loses in her urine, and the second that the older a woman becomes, the lower her bone density. When you put these two facts together, the logical conclusion might be that caffeine causes calcium to leech out of the bones, appearing in the urine as a waste product and ultimately causing osteoporosis.

This theory was supported by several studies, like one showing that women who drink more caffeine are at higher risk of hip fracture (Kiel 1990).

But on the flip side, there are many studies documenting that, while women are at higher risk for osteoporosis than men, their bone loss is independent of how much caffeine is ingested (Cooper 1992). Basically, bone loss happens in women—it's not caffeine's fault. A study of women in their twenties who drank moderate amounts of caffeine demonstrated that there was no bone loss and, in fact, a normal amount of bone gain.[2] There is also research suggesting that if there is a little bone demineralization as a result of caffeine, a glass of milk a day will offset it in an adult woman (Barrett-Connor 1994). So there has yet to be consensus on the subject of what caffeine does to our bones.

There is a negative side to caffeine that everyone does agree on: we develop tolerance to it. The body gets used to having caffeine around, so when you consume it regularly, it takes more in order to feel energized. A consequence of this is that regular caffeine consumers often go through withdrawal when they suddenly stop having it. The symptoms of caffeine withdrawal include headaches, irritability, depression, and in some cases muscle aches.

So how much caffeine are we talking here? Is a cup of coffee or a can of soda a little or a lot? The answer actually varies quite a bit. "Moderate" caffeine intake is considered to be 50–300 mg per day. That's a pretty big range. A typical five-ounce cup of coffee has between 71 and 220 mg of caffeine;[3] the same-sized cup of tea has 32–42 mg; and the same cup of hot chocolate has 4 mg. A twelve-ounce can of soda has 32–70 mg.

Worldwide, the average person consumes 76 mg of caffeine per day. But in the United States and Canada this number jumps to around 220 mg, and in Sweden and Finland people ingest more than 400 mg a day. (And almost all of it is from coffee!)

The range of reasonable caffeine consumption is wide. Down the street from my son's preschool there is a high-end coffeehouse. People seem to linger there for hours, endlessly sipping lattes. I grab my one cup of coffee and go. In the United States, adults who drink caffeinated beverages take in somewhere between 170 and 300 mg per day, with about two-thirds of it coming from coffee.

So what does any of this have to do with your kids? Just because they don't drink coffee, children aren't immune to caffeine exposure. In this group, caffeine comes from soft drinks (approximately 55 percent of caffeine ingested by kids every day), chocolate, and other beverages like energy drinks or tea. And by the teen years, coffee and coffee-based drinks like the Starbucks Frappuccino are par for the course.

Caffeine itself is not considered dangerous for children. It was once widely held that caffeine stunted growth, but this has been disproved in several studies. Aside from overdosing on caffeine pills—which is dangerous for anyone of any age—caffeine does not pose any true danger to our children.

There are, however, consequences to caffeine ingestion that do affect the health and well-being of kids. Sodas account for more than half of the caffeine consumed by children in the United States. They are relatively cheap, easily accessible, and coveted. Sodas are packed with sugar: one twelve-ounce can of regular soda contains about ten teaspoons of sugar. For this reason, soda consumption is associated with an increased risk of obesity. If a child drinks one soda every day, he increases his likelihood of obesity by 60 percent. To make things worse, obesity is associated with a series of other issues like diabetes and heart disease.

Cavities are also a bigger problem in soda consumers. Because of all that sugar, children who drink sodas are more likely to have cavities than those who don't.

Another consequence of caffeine ingestion is that caffeine seems to directly affect sleep, and for children (who require more sleep than their adult counterparts), this can take its toll. Caffeine makes it more difficult to fall asleep, causes a person to wake more often during the night, and decreases the amount of deep sleep. When you add up all of these things, caffeine drinkers get less total sleep each night. This, in turn, is thought to interfere with school performance and concentration. (Not to mention the anecdotal complaints from many parents of preteens and teens that sleep deprivation is directly correlated with moodiness!)

There are no published guidelines in this country that set a daily limit to how much caffeine a child should have. If we adopted Canada's recommendations, though, children and teens would be told to limit their caffeine intake to no more than 45 mg per day.

CAFFEINE CONTENT OF FOOD AND DRUGS
FROM CENTER FOR SCIENCE IN THE PUBLIC INTEREST

Coffees	Serving Size (oz.)	Caffeine (mg)
Coffee, generic brewed	8	133 (range: 102–200)
Starbucks brewed coffee	16 (Grande)	320
Dunkin' Donuts regular coffee	16	206
Starbucks Vanilla Latte	16 (Grande)	150
Coffee, generic decaf	8	5 (range: 3–12)
Starbucks Frappuccino	9.5	115
Espresso, generic	1	40 (range: 30–90)
Espresso decaffeinated	1	4
Teas	**Serving Size (oz.)**	**Caffeine (mg)**
Tea, brewed	8	53 (range: 40–120)
Starbucks Chai Tea Latte	16 (Grande)	100
Snapple iced tea (and diet)	16	10–42
AriZona iced tea, black	16	32
AriZona iced tea, green	16	15
Soft Drinks	**Serving Size (oz.)**	**Caffeine (mg)**
FDA official limit for soft drinks	12	71
Jolt Cola	12	72
Coca-Cola	12	35 (20 oz. = 58)
Diet Coke	12	47 (20 oz. = 78)
Mountain Dew, regular or diet	12	54 (20 oz. = 90)
Pepsi One	12	54 (20 oz. = 90)
Mello Yellow	12	53
Tab	12	46.5
Dr. Pepper, diet or regular	12	42–44 (20 oz. = 68)
Pepsi	12	38 (20 oz. = 63)
Diet Pepsi	12	36 (20 oz. = 60)
Barq's Root Beer, regular or diet	12	23 (20 oz. = 38)

Energy Drinks [4]	Serving Size (oz.)	Caffeine (mg)
Monster Energy	16	160
Full Throttle	16	144
Rip It, all varieties	8	100
Tab Energy	10.5	95
SoBe No Fear	8	83
Red Bull	8.3	80
Rockstar Energy Drink	8	80
SoBe Adrenaline Rush	8.3	79
Amp	8.4	74
Glacéau Vitamin Water Energy	20	50
Frozen Desserts	**Serving Size (oz.)**	**Caffeine (mg)**
Ben & Jerry's coffee ice cream	8	68
Häagen-Dazs coffee ice cream and/or frozen yogurt	8	58
Starbucks coffee ice cream	8	50–60
Chocolates/Candies/Other	**Serving Size**	**Caffeine (mg)**
Jolt Caffeinated Gum	1 stick	33
Hershey's Special Dark Chocolate	1.45 oz.	31
Hershey's chocolate bar	1.55 oz.	9
Hot cocoa	8 oz.	9 (range: 3–13)
Over-the-Counter Drugs	**Serving Size**	**Caffeine (mg)**
NoDoz (Maximum Strength)	1 tablet	200
Vivarin	1 tablet	200
Excedrin (Extra Strength)	2 tablets	130
Anacin (Maximum Strength)	2 tablets	64

Source: From http://www.cspinet.org/new/cafchart.htm, data collected September 2007.

WHAT IS THE BOTTOM LINE?

For your kids, make a big effort to go caffeine free. The reason to do so is not really to avoid the caffeine but rather to avoid the high sugar load and empty calories in caffeinated sodas and candy bars. A little bit in moderation is fine, but high intake—a soda per day—is associated with a significantly increased risk of obesity.

The biggest rumor about caffeine is untrue: it doesn't stunt growth. But excessive caffeine may make it difficult for a growing child to fall asleep, stay asleep, and get enough sleep. Restorative sleep is critical in order to have good energy during the day. In fact, some people theorize that one of the reasons children who drink caffeinated beverages are more likely to become overweight or obese is that they are more tired during the day and therefore less physically active.

For adults, a little caffeine is fine. God knows I adore my medium black coffee every morning. Moderate caffeine consumption is considered anything less than 300 mg per day. That's about one "Grande" (sixteen-ounce) coffee in Starbucks-speak. There is no clear data to support that drinking caffeine weakens bones or increases your risk for osteoporosis. Studies are published every year or two, but the results seem to consistently contradict each other.

If you decide to go decaf, it is worth noting that decaffeinated beverages aren't completely caffeine free. About 97 percent of the caffeine is removed in the process, but there are still 2–5 mg of caffeine in a cup of decaf (that's thirty times less than a regular cup of coffee) and just under 1 mg of caffeine in decaffeinated soft drinks.

WHAT'S IN MY HOUSE?

My house is soda free. We have it very occasionally for parties, but otherwise we don't. My kids have never really noticed—or if they have, neither has said anything. During my pediatrics training there was a time when I drank too much coffee. It just happens when you are up all night in the hospital. I cut back because all that caffeine made me feel crummy, not because I was worried about long-term health effects. But as I said before, and as any of my friends can tell you, I love my morning cup of coffee.

Chapter 9

Juice

WHAT IS THE QUESTION?

Fruit is a universal symbol of health and bounty. It comes from the earth, naturally sweet and sun ripened. Remember the ads from the 1970s and '80s about how to start your day off right? It was with cereal or toast, milk, and a glass of juice. Fruit juice was supposed to represent a glassful of health.

So what went wrong? Why is fruit juice on the receiving end of bad press, credited with everything from severe infections to steady weight gain? Is juice really all that bad for you? Is there anything redeeming—even healthy—about it?

WHAT IS THE DATA?

Fruit is a great source of fiber, vitamins, minerals, and water. Many fruits (especially dark berries like blueberries, grapes, and pomegranates) contain potent antioxidants that are thought to help repair damaged cells in the body. Cranberries are even credited with preventing urinary tract infections because they reduce the ability of bacteria to stick to the bladder wall. Cantaloupe is a great source of vitamins A and C, and bananas are full of vitamin B_6 and potassium.

When a piece of fruit is pureed, it retains all its nutrients in a blended mush. When it is juiced—that is, when the pulp is squeezed and the natural juices flow out—it still maintains much of its nutritional value.

This liquid, labeled as 100 percent juice, misses most of the fiber contained in the whole fruit. But otherwise, its composition is pretty close.

Much of the nutritional value of the juice is lost when 100 percent juice is diluted with other liquids, preserved for longer freshness, sweetened with extra sugars (or sugar alternatives), or altered in other ways in order to be conveniently packaged in a juice box. It has been transformed into a high-calorie, low-benefit drink.

The transformation of 100 percent juice into juice drinks has evolved over the past few decades. Advertising and labeling have made it increasingly difficult to tell which products are made from real fruit and which are packed with artificial flavors and colors. So when a parent wants to give his child a "healthy" drink, it is hard to figure out which drinks qualify. In fact, it has become downright difficult to determine how much juice is really in juice.

As a result of the juice being taken out of juice—or lots of other additives being added in—these drinks have become major sources of empty calories. Juice is routinely blamed for excess weight gain among American children, so parents are advised to avoid giving it regularly. The **AAP** recommends that kids drink no more than four to six ounces per day. While juice boxes are still staples at birthday parties (probably as a result of their convenient size and low cost), schools increasingly discourage them. Teachers don't want their students amped up on sugar. In schools teaching green living, the juice boxes are vilified as examples of trash generators because the straw wrapper, the straw, and the box itself are all pieces of garbage.

The reality is that juice in and of itself is not horrible for you. Juice purees are as good as taking a bite of the whole fruit. While 100 percent juices are more caloric than the actual fruit and lack fiber, they still provide essential vitamins and minerals.[1] Where juice falls apart is when it becomes dessert.

Tags like "cocktail," "punch," "drink," and "beverage" all mean that sugar has been added. When the label says "from concentrate," it may be fine, because all that has happened is that water has been removed and then added back to the juice. But read the label to make sure that lots of other additives—like extra sweeteners—haven't been put into the mix as well. This is why when you buy juice for your children, you should stick with 100 percent. You can feel comfortable that science is on your side, because recent data has shown that consumption of 100 percent fruit juice is not related to children becoming overweight (Nicklas 2008). Despite our desire to label smoothies as healthy, they are not. Smoothie stores may be designed to exude the look and feeling of health and fitness, but sixteen ounces of fresh fruit, ice, and ice cream is basically a milk shake, no matter how you package it. Simply put: anything other than real fruit, an all-fruit puree, or 100 percent juice generally isn't health food.

So the healthiest fruit juice is 100 percent juice. Does this mean that the fresh-squeezed products at the grocery store are best? Not necessarily. Though they contain all of the fruit's native vitamins and minerals and the least added sugar, they are often unpasteurized.

Pasteurization is a method of sterilization. Basically, liquids are

HOW MUCH JUICE IS IN VARIOUS TYPES OF "JUICES"

In the United States, the term "fruit juice" can only be used to describe 100 percent fruit juice. A blend of fruit juices with other ingredients (like high-fructose corn syrup) is called a "juice cocktail" or "juice drink." "Nectar" is a pure fruit or vegetable juice (or puree) diluted with water; it may contain sweeteners. It is confusing to read a label at a supermarket and know what you are getting. Here is a key:

"100 percent pure" or "100 percent juice": Guarantees only 100 percent fruit juice, complete with all its nutrients.

"Cocktail," "punch," "drink," or "beverage": Diluted juice (less than 100 percent), often with added sweeteners.

"Fresh squeezed juice": Squeezed from fresh fruit, not pasteurized unless noted on the label, and usually located in the produce or dairy section of the grocery store.

"From concentrate": Water is removed from whole juice to make concentrate; then water is added back to reconstitute to 100 percent juice or to make diluted juice such as lemonade.

"Not from concentrate": Juice that has never been concentrated.

"Fresh frozen": Freshly squeezed and packaged and frozen without pasteurization or further processing; usually sold in the frozen food section of the grocery store and ready to drink after thawing.

Juice on nonrefrigerated shelves: Shelf-stable product usually found with canned and bottled juices on nonrefrigerated shelves of your store; it is pasteurized or diluted juice, often from concentrate, packaged in sterilized containers.

Canned juice: Heated and sealed in cans to provide extended shelf life typically of more than one year.

heated so that bacteria and other sources of infection are killed off. According to the **FDA**, in the United States 98 percent of all fruit and vegetable juices are pasteurized.

There have always been people suspicious of pasteurization. Surely, they argue, heat destroys many of the beneficial components of certain drinks. Some of these people also contend that exposure to a small amount of bacteria is healthy, well tolerated by the intestine, and stimulating for the immune system.

Physicians, scientists, and researchers tend to disagree. Pasteurization prevents serious illness and even death. In 1996 a sixteen-month-old in Colorado died after drinking apple juice contaminated with the bacteria E. coli 0157:H7. During the same outbreak, sixty-six others were ill with the infection including a three-and-a-half-year-old who was hospitalized for twenty-four days.

E. coli 0157:H7 is the most famous bacteria—but not the only one—found in unpasteurized juices. It is been blamed for an average of twenty thousand cases of severe illness in the United States per year. While most people have heard of E. coli, they do not fully appreciate its ability to wreak havoc on the body. In fact, many physicians argue that because of widespread use of pasteurization, E. coli and other infections have become so rare that people have lost sight of their ferocity. But unpasteurized juices continue to sell, which poses a real health risk. Between 1993 and 1996, unpasteurized juices accounted for 76 percent of reported contamination cases, and illnesses associated with unpasteurized juices were more severe than those caused by pasteurized products.

WHAT IS THE BOTTOM LINE?

As with most things in life, the whole is better for you than the parts. If you are choosing between a piece of fresh fruit and a glass of juice, opt for the actual fruit. It has more fiber and fewer calories.

But when it comes to fruit drinks, here's where you want to make a real effort to find the healthiest alternative. Look for "100 percent juice" on the label and avoid anything labeled "beverage," "cocktail," "drink," or "punch." I advise drinking pasteurized juice to minimize the risk of consuming potentially dangerous bacteria. If your child is willing, though, I would always choose water or milk over any kind of juice.

I have taken care of many kids who do not drink milk. Either they have allergies or they have developed lactose intolerance or—in most cases—they simply don't like it. For such kids, 100 percent orange juice with calcium and vitamin D added is a good alternative. While OJ lacks the protein found in milk, it contains many of the other nutritional benefits.

Perhaps most important, don't pass off dessert drinks as healthy. Smoothies often provide a good load of vitamins and minerals, but they are drenched in ice cream or frozen yogurt or syrup sweeteners. Smoothies are treats, which are fine occasionally, but only in moderation.

WHAT'S IN MY HOUSE?

My kids love orange juice, so I keep it in the fridge. They drink a small glass about three or four times a week. My daughter discovered apple juice in kindergarten (I don't know why it took her so long) and asks for it all the time. Because most readily available forms of apple juice are juice drinks—they contain lots of sugar and very little nutritional value—it isn't all that different from a fruit roll or a cookie. So I simply don't keep apple juice in the house. I am sure my daughter drinks it at school, but at home I just gently remind her that our calcium-fortified orange juice is healthier, and that's why it's the juice available in our house. I figure that if I demonize apple juice too much, my daughter will continue to drink it out of rebellion. This way the novelty will wear off, and soon enough apple juice will go the way of princess paraphernalia and thumb sucking—things of the past.

Chapter 10

Milk

WHAT IS THE QUESTION?

Other than water, milk is often hailed as the best drink for kids and teens. Milk has loads of calcium to strengthen bones and provides protein to growing and moving muscles. It is part of a "healthy" breakfast, it remains a staple in grammar school lunchrooms, and it is the perfect partner to sinful cookies. Milk is credited with helping kids grow and even with reducing obesity among school-aged milk drinkers.

But lots of people see more bad than good in the milk carton. Milk is blamed for a variety of health problems, from congestion to constipation. Milk allergies are not uncommon these days. There is also tremendous concern about the dairy cows themselves because the foods cows are fed and the medicines cows take wind up in your cereal bowl. Unless you are drinking **organic** milk from cows not treated with **rBGH**, you may be getting a hefty dose of antibiotics or hormones or genetically modified organisms or pesticides, none of which very appetizing.

But what is the real deal? Is milk so important that its benefits outweigh its risks? How substantiated are the claims against milk? Are alternative milks—like goat, soy, and almond milk—better for you? Does milk really do a body good, or should you skip it altogether?

WHAT IS THE DATA?

The word "milk" can mean many things to many people. In its most standard form, we think of cow's milk, sold in cartons or bottles at just about every supermarket and convenience store across the country. But there is also goat milk, sheep milk, buffalo milk, soy milk, almond milk, coconut milk, rice milk, and breast milk, just to name a few.[1]

All female mammals produce milk. Humans have drunk the milk of other animals for thousands of years. Sheep and goats were domesticated around 8000 BC, and there is evidence that use of their milk followed not long after.

Cow's milk is naturally high in calcium. Other varieties of milk can also offer lots of calcium, but generally this is because the calcium is added supplementally. Calcium is famous for benefiting our bones, making them both long and strong. A single serving of milk (which is equal to one cup or 250 ml) has almost 300 mg of calcium. Milk is also a great source of protein. Its proteins come in two varieties. Casein accounts for about 80 percent, and whey accounts for the remaining 20 percent. In addition to providing muscle fuel, these proteins promote bone formation, slow bone resorption, and improve bone density.

Cow's milk contains other ingredients as well. Lactose—a precursor to the sugars glucose and galactose—gives milk its sweet taste. It also adds a hefty caloric load, accounting for more than a third of whole milk's calories. Milk has a variety of vitamins and minerals, including vitamins A, D, and K; the B vitamins (thiamine, B_{12}, riboflavin, biotin, pantothenic acid, and folic acid); iodine; potassium; magnesium; and selenium. It also contains Insulin-like growth factor-1 (IGF-1) that promotes bone mineralization as well as transforming growth factor-b2 that appears to slow adipocyte (fat cell) formation and may actually delay the onset of obesity.

The alternatives to cow's milk—soy, almond, goat, and so on—each contain a variety of these ingredients to a greater or lesser degree. Some have more proteins and some have fewer; some have greater amounts of vitamin A, while others have more potassium or magnesium; cow's milk has lactose, whereas soy milk, almond milk, and rice milk do not. But the point of this chapter is not to provide a nutrition lesson in milk ingredients. Rather, it is to answer two increasingly popular questions:

Do we need milk in the first place? And if so, is cow's milk necessary, or even safe?

Milk is the only nutrition a baby will receive for the first six months of life. Often it is given in the form of breast milk, though it may certainly come as formula or as a combination of the two. The first year of life is one time when cow's milk is uniformly considered unsafe, because it can cause allergic-type reactions. I call them "allergic-type" and not "allergic" because the number of children who are *actually* allergic to cow's milk is quite small. But the majority of babies in their first year of life will be sensitive to cow's milk: they might develop a rash around their mouth or bottom, have eczema on their skin, or have vomiting or diarrhea as a result of the exposure. In fact, many babies will have a teeny tiny bit of intestinal bleeding after drinking cow's milk—so little that a parent won't notice it, but enough that, if it goes on for a while, the child can develop iron-deficiency anemia.

Now, this gets really confusing because cow's milk is not okay for a baby but cow's milk formula is. The simple explanation goes like this: There are more proteins in cow's milk than in cow's milk formula; specifically, there's too much casein. The excess casein can irritate the lining of the intestinal wall, which is what can cause the symptoms of sensitivity like bleeding. But formula has fewer cow's milk proteins, with less casein, so it is easier for the infant gut to tolerate and digest. By the time a child is a year old, the intestinal wall can usually handle more casein protein. A one-year-old also has a more developed intestinal lining, is less likely to be sensitive to cow's milk proteins, and can absorb the minerals in cow's milk better. Coincidentally, this is the time than many parents switch from breast milk or formula to regular milk.

If this is the case, why do pediatricians recommend starting babies as young as eight or nine months on yogurt and cheese? While dairy products do have cow's milk proteins, it seems that if the dairy product is fermented the proteins are easier for the child to digest. Fermentation also breaks down the sugar lactose, creating smaller simple sugars that are more easily absorbed. Babies generally consume less yogurt and cheese than milk, so they can handle the amount of dairy protein coming their way in those forms. Much like cow's milk formula is okay, yogurt and cheese can be tolerated well before the first birthday.

Tolerated is one thing; necessary is another. Why do most nutritionists and pediatricians emphasize the importance of milk? Because milk helps children grow. It is a cheap, easily accessible source of protein and calcium. Milk drinkers tend to be taller than non-milk drinkers, and they definitely have better height gain in puberty (Okada 2004). Many people attribute this to the calcium load in milk, though there is evidence that milk may also directly increase the level of growth hormone produced by the body (Rich-Edwards 2007).

The benefits of milk even extend into adulthood. Low-fat milk consumption appears to decrease hypertension, coronary artery disease, and colorectal cancer. Among overweight people, it may reduce the likelihood of insulin resistance and type 2 diabetes. And compared to non-milk drinkers, milk drinkers have a lower risk of bone fracture (Konstantynowicz 2007).

Perhaps the biggest benefit of milk is its association with lower body mass index (BMI). Multiple studies have demonstrated that, compared with those who don't, children and teenagers who drink milk are less likely to be overweight. It doesn't seem to matter whether a child drinks plain milk or chocolate milk or strawberry milk—all flavors are associated with improved nutrition and no increase in BMI (Murphy 2008). But some of these studies are a little misleading, particularly when comparing milk drinkers with soda and juice drinkers. Soda and juice are well-known sources of empty calories. The studies were designed this way because it is the rare child who is willing to drink water and only water, so the idea was to compare choices other than water. These results prove that milk is a much healthier water alternative than juice or soda.

Despite the benefits of milk, there are many who believe the negatives outweigh the positives. The litany of complaints against cow's milk is long. Here is a list of the most common among them. They are not in order of importance, but, rather, I begin with the simplest concerns and move toward the more complex.

First, cow's milk causes constipation. This is true. Cow's milk is clearly associated with constipation, particularly among toddlers. The constipation is made worse in this population by functional withholding (it feels bad to have a constipated bowel movement, therefore these kids often hold it in, making the constipation worse). Typically, the constipation resolves when cow's milk is stopped, and, interestingly, it often does not return if the milk is restarted.

Second, cow's milk causes diarrhea. Yes, there are complaints in both directions here. The diarrhea caused by cow's milk is by and large a function of age. Among the little kids (infants and toddlers), diarrhea is from milk sensitivity; starting around age five, it is a lactose-intolerance issue. Among infants and toddlers, about 90 percent will outgrow this sensitivity and tolerate milk just fine by the time they go to preschool. The older lactose-intolerant crowd, though, generally needs to find a non-cow form of milk or to drink Lactaid, which provides the enzyme (lactase) necessary to break down lactose.

Third, cow's milk can be an allergen. Some—though clearly the minority of—people who have diarrhea after drinking cow's milk are actually allergic to it. These folks almost always have other allergic symptoms too. More than half have eczema, which is dry, irritated skin that can appear anywhere but in children is most typically found in the skin folds (elbows, knees, wrists, ankles) and on the cheeks. People allergic to cow's milk can develop hives or wheezing or vomiting. Allergic reactions to milk can be profound, no doubt, but so too can allergic reactions to any food or drink. Avoiding cow's milk products has never been proven to prevent the eventual formation of milk allergy. If you avoid giving your kids milk early on, all you do is prevent symptoms of allergy. If your kid is going to be allergic to cow's milk, he will be allergic regardless of whether you expose him when he is young.

Fourth, cow's milk causes nasal congestion. There isn't a lot of data supporting this, but anecdotally I find it to be true. I suspect that this group of patients has a subclinical allergy—an allergic reaction to milk that has never really caused significant problems (no rash, no diarrhea or vomiting) but does cause inflammation in the nose and sinuses.

Fifth, cow's milk is associated with acne (Adebamowo 2006). There are theories that this has to do with the hormones given to some cows to stimulate their milk production (more on this to follow), but it is not entirely clear. The acne may also be caused by the IGF-1 normally produced by all cows, whether they are living organically or injected with hormones.

Sixth, some people believe that casein protein in milk is associated with autism. They argue that the body breaks down casein into casomorphin and this compound triggers autistic features. Very few studies have looked at this question, and there is no data to clearly support this theory. But that said, many parents with children on the autistic spectrum choose to give their children gluten-free, casein-free diets and report behavioral improvements. As long as those children continue to eat balanced diets and get enough protein, I see no downside to trying it.

Seventh, cow's milk causes symptoms of lactose intolerance. For the record, lactose intolerance is not the same as milk allergy. A person who is lactose intolerant has lost the ability to digest milk because they no longer have enough lactase, the enzyme that breaks down lactose. The majority of adults in the world are lactose intolerant—milk causes them to have cramping or gas or diarrhea. Lactase levels usually peak during infancy, and then they slowly decline unless milk is consumed regularly (and even still they may decline). It is not uncommon to see lactose intolerance begin to appear around kindergarten age (five or six years).

Eighth, there may be links between milk and adult diseases. One study demonstrates an association between milk (and general dairy) consumption and Parkinson's disease, though the association is only seen in men (Chen 2007). Another study correlates high milk intake and prostate cancer (Chan 2004). Certainly, more associations will follow. But it remains to be seen whether these claims can be substantiated by repeat studies. If the association between milk and acne turns out to be a hormone effect, it is then feasible that there will also be a connection between milk and hormone-dependent cancers.

The list goes on, though for each new claim there are fewer and fewer studies, providing less and less evidence. In fact, as I wrote this particular chapter, I happened to be sitting in a coffeehouse filled with screenwriters (I am in LA, after all). The men sitting next to me were taking a writing break, educating each other about the myths of milk, particularly the calcium claims. One man had recently become vegan, and he had a lot to say about the evils of dairy. While some of his information was good (kale is a better calcium source than spinach because it has less oxalate, so its calcium is better absorbed), most of it was downright wrong (like that our gut absorbs less than 5 percent of the calcium in cow's milk). Eavesdropping on their conversation reminded me that most people get their medical information from their friends, who are often inaccurate or just plain wrong.

But back to the question at hand. The list of complaints about milk is long, and the concerns are not unfounded. The milk fans and the milk critics seem to balance each other out. That is, until you add to this a newer variable: growth hormones given to dairy cows. Independent of all the concerns about milk itself, hormone treatment adds a heavy layer of health implications.

A huge number of dairy cows—somewhere around 35 or 40 percent—are given rBGH, short for "recombinant bovine growth hormone." [2] The rBGH increases a cow's milk production by about 10 percent. It's just a lab-manufactured version of the growth hormone that is naturally produced by cows. The FDA approved rBGH in 1993, and it is sold under the trade name Posilac. Today it is used in almost every state (Michigan is the sole exception) and by at least eight thousand dairy producers at last count. It is the top-selling pharmaceutical product for dairy cows in the United States.

There are two separate concerns about rBGH: what it does to the cow and what it can do to the human milk drinker.

Let's start with the cow. Many reports suggest that rBGH-treated cows are sicker than nontreated cows. They are more likely to have symptoms of lameness, such as foot problems. They also can have reactions at the site of rBGH injections.

It is well known in the dairy industry that rBGH-treated cows are more likely to get mastitis, an infection of the mammary glands (which are the milk-producing sacs and ducts). Mastitis is a bacterial infection, so there are concerns that dairy cows with frequent mastitis may produce milk infected with bacteria. On top of this, to treat a bacterial infection the cow must receive antibiotics. It wouldn't be a major concern if a dairy cow were to get antibiotics once or even twice. But if a cow is getting mastitis frequently, she is going to receive antibiotics just as often. If those antibiotics pass into the milk, human consumers are then exposed to them. Ultimately, the antibiotics could cause a milk drinker to have an allergic reaction, or they can simply linger in our own bodies, leading to antibiotic resistance.[3]

Separate from worries about dairy cows, many people have concerns about the direct effects of rBGH on humans. If we drink milk from rBGH-treated cows, and we ingest the rBGH, what does the hormone do to our bodies? The FDA and the NIH have both published statements that rBGH does not directly affect human health. While there have been claims that the artificial hormone may have effects on the human immune system (for instance, some contend that it triggers allergic reactions), this is not really supported by science. This is because rBGH itself does not pass through to the milk we drink in significant quantities.

On the other hand, the milk from rBGH-treated cows does have significantly—albeit slightly—higher IGF-1 levels. IGF-1 has been implicated in breast cancer, prostate cancer, and colorectal cancer. So do rBGH-treated cows produce milk with enough extra IGF-1 to increase our cancer risk? This question has not yet been answered.

As a result of all of the debate over rBGH, and with a great deal of science in the background, several countries have banned the synthetic hormone. Canada, Australia, New Zealand, Japan, and the European Union do not allow treatment of their dairy cows. In the United States, several large retailers—such as Kroger, Wal-Mart, and Costco—have recently made moves to get milk from rBGH-treated cows off their shelves. Even the country's largest dairy processor, Dean Foods, no longer sells milk from cows receiving rBGH.

Some people have made a quick jump away from cow's milk because of the rBGH controversy. They figure that choosing an alternate milk source is safer than not knowing whether the milk at your local coffee joint has rBGH.[4] Soy milk is the most common default milk because it is readily available. But is it actually safer?

The quick answer is no. While organic soy milk is a healthy cow's milk alternative, nonorganic soy is chock-full of **GMOs**, or genetically modified organisms. No one has done a study determining which is worse for you: rBGH-treated cow's milk or GMO-affected soy, but it seems pretty logical that neither is ideal.[5]

The healthiest—and safest—type of milk is organic: free of chemical herbicides, pesticides, GMOs, antibiotics, and hormones. Organic soy is grown with more natural fertilizers, and organic cows generally have a higher standard of living, grazing over larger patches of land for more hours of the day than their nonorganic cousins. So choosing organic milk—be it almond or rice or soy or goat or cow or buffalo—is probably the most important variable in the equation.

WHAT IS THE BOTTOM LINE?

When my daughter was a year old, she switched to cow's milk and became constipated. For a couple of years, she drank organic soy milk and did quite well. Just as the textbooks said, around the age of three a magical switch flipped and her issues resolved. She alternated between cow and soy milk for a while, but now prefers cow's milk, so that's what I give her.

I have no problem with families choosing alternative milks. I do, however, believe that the calcium and protein load in milk is hard to replace with other foods and drinks. So I think some form of milk—animal, nut, soy, or rice based—is better than none.

For my patients who are truly milk allergic or who just cannot tolerate the taste of milk, I think calcium-fortified orange juice is better than nothing. But when I see kids in the office year in and year out, my experience is consistent with what the studies say: the milk drinkers are taller and often leaner.

Beware of all dairy products and not just milk. Cheese, yogurt, cottage cheese, and so on come from the same dairy cows as milk. Buying organic yogurt or cheese is just as important as buying organic milk.

It seems to me that the food we eat and the liquids we drink are increasingly modified without our express knowledge. Milk is the smallest offender here: non-rBGH (or non-rBST) is prominently advertised on the label. But we ought to be fighting for more labeling in the future. Consumers deserve to know what they are eating and drinking, and if they aren't sure what the acronyms stand for, they should educate themselves. There have been many moves within the dairy industry to remove rBGH labeling from milk cartons, and public health advocates have prevailed. Perhaps we should make note of this and apply the lesson to other foods, like GMOs.

WHAT'S IN MY HOUSE?

My children drink cow's milk every day. I also drink it every day, as does my husband. I buy organic, and this is one item I will not skimp on.

Chapter 11

Sports Drinks, Vitamin Waters, and Energy Drinks

WHAT IS THE QUESTION?

The beverage shelf at the market used to be limited to soda and juice. Then bottled water came into vogue in the 1980s. When I was in high school, the thing to do was to carry around your own water bottle, usually an Arrowhead or Evian container refilled at the school drinking fountain.

Over the past quarter of a century, the field of bottled beverages has taken on a pharmaceutical bent. Now store-bought drinks don't just promise to quench thirst but also to wake you up, enhance your physical performance, and even fill your nutritional gaps.

Energy drinks, vitamin waters, and sports rehydration solutions make promises of health, immunity, and stamina. Are these claims true? Some people—including my own brother—think vitamin waters are a reasonable substitute for balanced nutrition. Are these specialty drinks worth the added cost? Are they safe for kids? And what's wrong with plain old water?

WHAT IS THE DATA?

This collection of drinks—sports drinks, vitamin waters, and energy drinks—doesn't really belong together. Sports rehydration solutions like Gatorade are marketed to athletes who want to replenish their hard-earned sweat with an electrolyte-balanced solution. Vitamin waters are waters treated with health-enhancing nutrients like vitamins and minerals and usually a splash of refreshing flavor. And energy drinks are sugary and caffeinated alternatives to coffee, offering a pick-you-up in the form of a juice or soda. I only lump them together because they represent a new generation of beverages that promise to do more than just satisfy your thirst.

Each of these drinks offers a **dietary supplement**: either something you should be eating or something you might need to replenish. Some of these drinks claim to treat any number of physical problems. The ingredient list reads like a pharmacopoeia, with vitamins (starting with A and going up through the alphabet), minerals (like calcium, magnesium, and phosphorus), herbs (like Saint-John's-wort, ginkgo, and ginseng), and stimulants (caffeine, taurine, occasionally even ephedrine).

There is another thing all these drinks have in common, and that's sugar. Sometimes it comes in the form of sucrose (essentially regular table sugar), sometimes as high-fructose corn syrup, and in the diet-varieties—of which there are surprisingly few—artificial sweeteners like aspartame are used.

Many people are surprised to learn that these drinks are anything less than pristine, let alone fattening. Vitamin water sounds so natural and healthy. Sports rehydration drinks replace your sweat with balanced fluids. Energy drinks are supposed to rev you up so you can get more accomplished, not weigh your body down with calories. One thing's for sure, though: all of these drinks are sweet. And the ones with sucrose or corn-syrup sweeteners are sweet at a caloric price. Somehow the marketing has won out (especially with vitamin waters), and consumers tend to overlook the downsides to these beverages.

How did we get to this point in the first place? What's wrong with plain old water? Gatorade was the first trendsetter, back in the late 1960s. As most people know, it was designed for the University of Florida football team, the Gators. The coach noticed that players seemed to have less energy in the heat, even when they drank plenty of water. What he had recognized was that when athletes sweat they don't just lose water. Sweat is salty. So the team doctor and a group of researchers mixed water with sodium, potassium, and phosphate in order to replace the electrolytes lost in sweat. They also threw in lemon juice and some sugar. Voilà! Gatorade was born. When the Gators won a big bowl game in 1967, they credited their signature drink. The rest is history. Now Gatorade comes in a number of varieties, including high-sodium and high-potassium versions designed for intense athletes, shakes, and even nutrition bars. Today there are lots of sports drinks available. They all contain some combination of sugar and electrolytes. Across the board, the sugar load is pretty hefty, designed to provide immediate energy. Gatorade has 14 grams of sugar per serving. A serving size is only eight ounces, but most Gatorade bottles are actually twenty ounces, so they have 2.5 servings. That means that when you read the ingredient list, unless you plan to stop drinking at eight ounces (and most people don't—most drink the whole thing), you need to multiply the calories and the electrolyte contents by 2.5 to figure out how much you are getting. If you drink all 35 grams of sugar in a twenty-ounce Gatorade, that's 125 calories. Compare that to a can of Coke with 39 grams of sugar and 140 calories—it's not so different. The point here is that sports rehydration drinks tend to have a lot of calories. If you aren't exercising aggressively, you shouldn't be drinking a sports drink.[1]

Almost all sports drinks contain sodium and potassium because these are the principle electrolytes in sweat. But sweat varies quite a bit from person to person. Some people sweat a lot, some very little. Some people have very salty sweat packed with sodium while others have watery sweat. Most people do not need a special sports drink—there is no need to replace their sodium and potassium immediately after exercising because the body does a good job of maintaining electrolyte balance. There was lots of sweating before there was ever Gatorade, and people survived.

On the other hand, sodium and potassium have been hailed as the key ingredients in sports drinks. This was bolstered by studies in the 1970s and '80s showing that people who exercised and then drank fluids containing electrolytes maintained better plasma volume. In other words, they were rehydrated more effectively (Costill 1973; Nielsen 1986). Only sodium and potassium have ever been proven to help in this regard; other electrolytes don't make a difference. But again, most people don't need a sports drink after exercise. The group that benefits the most are those that work out aggressively for more than an hour at a time. This is because by the end of the workout, they are a little bit dehydrated.

I don't typically recommend sports drinks for young garden-variety athletes, but I do find them very helpful when kids are vomiting or having diarrhea. It goes to the dehydration issue. The sodium and potassium in sports drinks help turn around dehydration a little faster. There is also a benefit to the sugar—when a child has a stomach virus, he will throw up and have diarrhea, and usually he doesn't want to eat. So his blood sugar drifts down, making him moody, at first, then downright sleepy. The sugar in sports drinks is a fast-acting pick-me-up. Kids feel better—at least better enough to get up off the couch and maybe even try a piece of toast.

Manufacturers have caught on to the fact that doctors like to use sports drinks to help rehydrate patients. So they have marketed some as if they were medicines. In the field of pediatrics, Pedialyte is the classic example. This is essentially Gatorade without the sugar. It has sodium and potassium dosed appropriately for kids. The intentions here are quite good: it is sugar free because in some cases sugar can actually worsen diarrhea. But have you ever tasted Pedialyte? Most flavors of the drink are unpalatable—at least according to many pediatric patients. So I either suggest sports drinks or Pedialyte Freezer Pops, which have a remarkably good taste even though their parent drink doesn't.[2]

It's a small leap from sports drinks to vitamin waters. If athletes could replace the sodium and potassium lost in sweat, couldn't the average Joe replace vitamins and minerals that he just doesn't happen to get into his regular diet? The answer is yes. So over the past decade, along came vitamin waters.

Vitamin waters contain a variety of vitamins and minerals. Each brand is different. In fact, each flavor within a brand is different. Most contain vitamins B and C, which are water-soluble vitamins. This means that if you drink them in vitamin water and on an empty stomach, there is a good chance the vitamins will be absorbed by your body. Water-soluble vitamins get in and out of the body fairly easily, and they don't require food. But some vitamin waters contain the fat-soluble vitamins A, D, E, or K. These vitamins require food for absorption, so if you are drinking them on an empty stomach, they are likely going straight through you.

Vitamin waters sound too good to be true, and they are. The first—and biggest—problem is the sugar load. Walk down the aisle at the market and you will see the bright colors of various vitamin waters shining at you: fuchsia, aquamarine, emerald green. Their colors are often matched with sugar-infused flavors. A typical vitamin water has 13 grams of sugar per serving, and, as with sports drinks, there are 2.5 servings in a bottle. So if you drink the whole thing, you are in for 32.5 grams of sugar (that's about 120 calories). Compared to a can of Coke that has 39 grams (140 calories), there's not a big difference.

The next issue is one of setting expectations: kids who get used to having vitamin water think water should taste sweet. I am not exaggerating. Many school-aged kids have told me that they don't like water anymore because it doesn't taste like anything. Vitamin waters, on the other hand, are fruity or tangy and, as a result, kids prefer them. With the word "water" in the title, parents often don't see the harm, but if they were a bit more aptly named (like "sugar liquid"), parents might pause and think twice before buying.[3]

Vitamin waters can have other things thrown in as well, all advertised as health boosters but many not proven to help you. Often these drinks include herbal supplements or caffeine. These additives have no documented benefit for kids or haven't been studied in children in the first place. Ironically, even the vitamins provided in vitamin water have no proven benefit. Most doctors and scientists say that if you want to supplement your child's vitamin and mineral load do it with a multivitamin once a day—they are better studied, cheaper, and won't cause weight gain. If you do give your child a multivitamin, don't double up with vitamin water. Believe it or not, you can overdo it with certain vitamins.

It is possible that, with all of these bonus ingredients, vitamin waters can actually cause harm. For people who take medications every day, the additives in your vitamin water may interact with your drugs. Doctors are often careful to warn about well-known drug interactions—for instance, don't drink grapefruit juice if you are taking a cholesterol-lowering statin—but most patients don't think to disclose vitamin water to their doctors. Even if patients do, each brand has a unique combination of ingredients, making it almost impossible for doctors to forewarn patients appropriately.

Overall, there aren't many people who think vitamin water is dangerous. It's just unnecessary and full of calories. So if you are thinking of buying it because it will make a big impact on your health, keep on walking down the aisle. If your kids have gotten used to the taste of it, spike their regular water with a splash—and I mean just a splash—of juice.

On the grocery shelves next to the vitamin waters (meant to connote health) sit the energy drinks (symbols of enhanced power and strength). The first energy drink, called Lipovitan D, was marketed in Japan in the early 1960s. It took twenty years for an energy drink to come to the United States, and when it arrived it was in the form of Jolt Cola. Jolt was actually marketed as a soda that provided an extra energy boost. In this country, it took a few more years for energy drinks to move away from the realm of sodas, but today there is a wide variety of energy drinks available.

Energy drinks are a huge business: it is estimated that Americans spend $750 million per year on this group of drinks alone. The most well known is Red Bull Energy Drink, first distributed in 1997. For more than ten years, Red Bull has led the field.

Red Bull is fairly typical of energy drinks. Its primary ingredients are caffeine, glucuronolactone, and taurine. Caffeine, of course, is a stimulant. Glucuronolactone is a sugar that comes from glucose. Taurine is a chemical naturally produced by the body: it has a variety of biological jobs, including playing a role in skeletal muscle contraction. No one is exactly sure what taurine does in energy drinks, but it appears on the ingredient lists of many of them.

Red Bull gained instant popularity on the party circuit because it keeps people awake for hours. There are many who swear they use it innocently, not to party but to study through the night. Its boost is nothing magic—it is simply from the caffeine and the sugar. The caffeine load can be surprisingly high, with one can of Red Bull containing as much as two or three cups of coffee, and this has generated concern. Caffeine acts as a diuretic, so rather than rehydrating the drinker, it actually *de*hydrates him. This plus the presence of various stimulants can cause—and has caused—fatalities. Red Bull is also popular on the nightclub scene, where it is often dangerously mixed with alcohol and other drugs. For this reason, Red Bull is banned in France, Denmark, and Norway.

Another criticism of Red Bull is its effect on teeth. Like most energy drinks, Red Bull is highly acidic. The acids are known to cause significant erosion of tooth enamel.

And finally—here we go again—Red Bull has a huge sugar load. One can has the same amount of sugar, 39 grams, as a can of Coke. Red Bull actually has more calories than Coke, at 160 per can. Despite this unflattering information, there are many energy drink advocates.

Some argue that energy drinks are a better source of caffeine than coffee because the dosing is consistent—you know how much you are getting in each can of Red Bull, while you don't really know how much is in a given cup of coffee. Data has actually backed this up: in one 2002 study, energy drinks consistently improved the alertness of sleepy drivers (Reyner 2002). There are also studies that show energy drinks increase aerobic endurance and mental performance (Alford 2001). But while this may be true, any boost from an energy drink is really just a function of the sugar and caffeine.

Energy drinks other than Red Bull advertise their own unique list of ingredients. A very few contain ephedrine, a stimulant. Others have ginseng (a medicinal root thought to reduce stress and boost energy), guarana (a stimulant found in a South American shrub), or ginkgo biloba (a presumed memory enhancer). Some energy drinks have the amino acid carnitine, others have the organic acid creatine, and still others have members of the B vitamin complex.

There is very little evidence that any of these ingredients do anything they claim to. With the exception of drinks containing ephedrine, there is also very little evidence that these ingredients are dangerous. There have been reports of seizures in people who drank high volumes of energy drinks, but the drinks were never proven to be the cause.

WHAT IS THE BOTTOM LINE?

Specialty drinks like sports rehydration solutions, vitamin waters, and energy drinks don't hold any magical keys to health. In most instances, plain old water does a body just as much good. In fact, regular water is usually better because it doesn't have any calories.

Sports drinks have electrolytes that benefit intense athletes. They also tend to help people who are slightly dehydrated from vomiting and diarrhea. But they have lots of sugar, so unless you are exercising a lot and burning through those calories, they can be fattening. In the case of diarrhea, the sugar in drinks like Gatorade can sometimes exacerbate the problem.

Vitamin waters are an expensive and highly caloric alternative to a multivitamin. Most of the vitamins they offer are potentially helpful but usually unnecessary. A few (like the fat-soluble vitamins A, D, E, and K) go right through you unless you are drinking the vitamin water with a meal. The problem with these drinks is their sugar load, which is not all that different from soda's.

Energy drinks sell the promise of power, alertness, and productivity. But really what they have is a load of caffeine and sugar. Sure, there may be a variety of other ingredients making claims—like ginkgo for memory—but these have no proven benefit. Basically, energy drinks are not much different from sodas minus the carbonation.

So stick with regular old water, even (gasp!) tap water. It's free: it doesn't cost anything, it has no sugar (so it won't pack on the pounds), and it has no caffeine.

WHAT'S IN MY HOUSE?

Anyone who comes over to my house knows that the fridge is boring: there's milk, orange juice, and water. My husband has mastered the homemade smoothie (OJ, yogurt, and ice), and that's what my kids ask for in lieu of vitamin water. I actually get upset when I see young kids carting around enormous bottles of vitamin water, because I don't think their parents realize how caloric it is, with no real health benefit.

Part III

In Your Surroundings

Chapter 12

Cell Phones and Electromagnetic Radiation

WHAT IS THE QUESTION?

What was an uncommon accessory only a dozen years ago is now virtually ubiquitous among adults, teenagers, and even tweens. An estimated 265 million Americans currently have cell phones. This is just the tip of the worldwide iceberg: globally there are three billion cell phone owners.

Cell phones emit electromagnetic radiation. This isn't cause for panic—all objects with a temperature above absolute zero emit radiation, everything from the human body to the sun. And anything that has a battery or plugs into the wall generates electromagnetic radiation, including all wired and wireless technology.

Existing safety guidelines for all electronic and wireless devices were developed in order to avoid two known dangers to the body: excessive heat and specific electric currents called radio frequency. Every object in this group, from power lines and microwave ovens to cell phones and laptop computers, has been designed to keep the heat minimal and the frequency of the waves limited to a specific part of the electromagnetic spectrum. These safety standards were meant to mitigate cell damage in the human body.

But now many experts believe that these standards don't make technologies like cell phones safe enough. It appears that previously underestimated nonthermal and low-frequency electromagnetic radiation actually does affect cells, evidenced by cell mutations seen in the research lab. This creates a whole new issue, because no one knows what this research means. Do cells exposed to certain electromagnetic fields mutate in the body like they do in the lab? Does this mean that electromagnetic radiation causes brain cancer? Studies show that our cells are sensitive to exposures hundreds of times below the federally established levels for public safety. Does this mean that the radio frequency emitted by a cell phone, long considered safe, is actually worrisome?

The debate now is no longer whether cell phone (and other wireless-generated) radiation affects the body but rather *how*. Are the devices we use every day dangerous? If so, exactly what kind of danger are we talking about? And when it comes to a cell phone, are you any better off if you use an earpiece rather than hold the phone up to your head?

WHAT IS THE DATA?

Understanding electromagnetic radiation in general and cell phones in particular can be exceptionally confusing. I think it is important to state this at the outset, because the debate over wireless technology is one of those evolving topics with lots of big terms thrown around, many of which have scary connotations. Just take the phrase "electromagnetic radiation." Many people hear the word "radiation" and immediately conjure images of Hiroshima—it has become a scary, dangerous, negative word. But radiation isn't automatically frightening or harmful.

We live in a radiation-filled world. The earth emits electromagnetic radiation and has since its inception. Our brains are actually electric organs. The brain uses electric signaling between cells to communicate quickly. The heart, too, is an electric organ, relying upon its own internal pacemaker to tell it when to beat. If our bodies did not invoke electric signals, they would not work nearly as efficiently and quickly as they do.

In modernizing our world, humans have built devices that use electricity and batteries. Anything that is "powered" emits radiation of some sort. There is no person alive on the planet today who was born *before* electric and battery-powered devices were conceived.

Electromagnetic radiation is all around us every day. This includes (but is certainly not limited to) TV and radio signals, power lines, car engine spark plugs, microwave ovens, cordless phones, computers, walkie-talkies, washers, dryers, refrigerators, garage door openers, remote controls, and, of course, cell phones.

So many people have cell phones that if there is a health consequence—even in a teeny percentage of people—it will translate into millions of sick people. For example: if cell phones were found to cause harm to 1 percent of all users, 2.6 million Americans would be affected. The point here is that since the number of people using cell phones is so massive, the safety standards should be even more stringent.

In order to understand the debate over what effects a cell phone may have on the human body, it helps to have a quick refresher course in physics and a handful of definitions. When a current is supplied to an antenna, it generates an electromagnetic field (EMF) that travels through space. The field is made of wavelengths. Picture a wavelength as an invisible wavy line traveling through the air. Some lines have long, slow waveforms while others have many short, fast waves. Both are types of EMFs with different characteristics. Radio frequency (RF) refers to a specific range of wavelengths that is above the "radio" range (10 kHz) and below infrared light (300 GHz). In other words, RF represents a limited range of frequencies within the electromagnetic spectrum—it is a type of EMF. All wireless transmissions use RF including AM and FM radio, TV, satellites, portable phones, cell phones, and wireless networks. Wireless devices also emit EMFs. Cell and cordless phones, cell towers, and broadcast towers all generate EMFs.

The term "nonionizing" is also thrown around. Nonionizing radiation is generally defined as any type of electromagnetic radiation that does not carry enough energy to ionize atoms or molecules. In the past, nonionizing radiation was considered to be unable to cause DNA mutations or cancer. RF is technically nonionizing radiation.

Now, let's apply the technical terms to cell phones specifically. A cell phone emits EMF from its antenna. Imagine the field is a series of concentric circles radiating from the antenna—the closer the circle is to the antenna, the more intense the field. This is important because if you hold a cell phone up to your head, your brain becomes part of the electromagnetic field; if you have an earpiece in your ear and the earpiece is attached, by a rubberized wire, to a cell phone that is sitting two feet away from your body, your brain is not within the field. Or said more precisely, your brain is still within the field, but since you are farther from the antenna,

The Electromagnetic Spectrum

Chart by LASP/University of Colorado, Boulder

Wavelength		Band		Visible
10^{-6} nm				
10^{-5} nm				
10^{-4} nm		Gamma-Rays		
10^{-3} nm				
10^{-2} nm	1 Å			
10^{-1} nm				
1 nm		X-Rays		Violet
10 nm				Indigo
100 nm		Ultraviolet		Blue
10^3 nm	1 μm	Visible Light	Visible Light: ~400 nm – ~700 nm	Green
10 μm		Near Infrared		Yellow
100 μm		Far Infrared		Orange
1000 μm	1 mm			Red
10 mm	1 cm			
10 cm		Microwave		
100 cm	1 m		UHF	
10 m			VHF	
100 m			HF	
1000 m	1 km		MF	
10 km		Radio	LF	
100 km				
1 Mm			Audio	
10 Mm				
100 Mm				

nm=nanometer, Å=angstrom, μm=micrometer, mm=millimeter,
cm=centimeter, m=meter, km=kilometer, Mm=Megameter

the field that your brain sits in is much weaker at that point. In fact, according to Ronald Herberman, director of the University of Pittsburgh Cancer Institute, when the antenna is just two inches away from the head, the strength of the EMF drops by 75 percent.

There are thousands of bands within the RF range, each of which corresponds to a specific device or luxury of modern life. Baby monitors use 49 MHz, garage doors use 300–400 MHz, cell phones use 800–850 MHz and 2400 MHz, and GPS devices use 1227 and 1575 MHz.[1] Therefore, in our daily lives we come into contact with hundreds of devices that utilize RF and emit EMFs.

People tend to focus on cell phones, but cordless phones have the same anatomy. When you walk around your house chatting on the cordless phone, the antenna is held next to your head and, as with a cell phone, your brain is close to the source of the EMF. Therefore, in considering safety issues, similar rules apply to both cordless home phones and cell phones.

More than a dozen studies have directly linked EMF from cell phones and cordless phones with brain tumors and acoustic neuromas. The latter are benign tumors of the acoustic nerve between the ear and the brain. One group of researchers, led by Lennart Hardell, has published several studies on the topic in Europe. One study claims that among people who have used a cell phone for more than ten years and who hold the phone to alternate sides of the head, listening with both the right and left ear, there is a 20 percent increased risk of brain tumor; for people who hold the phone only on one side, there is a 200 percent increased risk for tumor. The same researchers suggest that cordless home phones are worse: holding the phone on alternate sides has been associated with a 220 percent increased risk of brain tumor, while holding it on one side is associated with a 470 percent increased risk (Sage 2007).

Other cancers have been linked with EMFs too. The strongest association is between EMFs and childhood leukemia. There are also associations between long-term EMF exposure and breast cancer, EMFs and prostate cancer, and EMFs and adult leukemia. Noncancerous diseases have also been tied to EMFs, including Alzheimer's disease and some types of seizure disorders. All this data is based on small or retrospective studies and is often dismissed by critics as not convincing.

The studies linking cancer with cell phone use are constantly in question. The general rule in science is that in order for a theory to become accepted knowledge, the data must be reproduced over and over again. For this reason, while there is good data showing a connection between cell phones and cancer, there is also a strong argument against it: the data linking EMFs and cancer has not been consistently reproduced. As a result, a sizable group of researchers argues that this means cell phone use does not cause cancer, and therefore cell phones are safe. In medicine and science, if you can't prove something over and over again, it isn't necessarily true.

The scientists who believe that EMFs and brain cancers *are* linked argue that with 265 million cell phone users in this country alone (and growing), it is increasingly impossible to reproduce studies because there is a rapidly diminishing control group. Essentially, there are fewer and fewer people with zero exposure to cell phones or EMFs. They also say that the studies documenting cell phone safety are too short, most lasting three years and a few up to ten. Once longer-term data is available, these scientists believe the association between EMFs and cancer will be irrefutable. They worry that, while there seems to be good preliminary science supporting a connection between cell phones (and EMFs in general) and cancer, it may be a race against time to definitively prove causation.

If cell phones do cause cancer, we need to be able to explain how these devices actually do damage. Unfortunately, no one in the scientific community agrees on this either. Remember that nonionizing radiation wasn't supposed to cause cell mutation or cancer. So people are busy coming up with theories about how EMF from a cell phone leads to bodily harm. Their ideas involve everything from the immune system to free radicals to dividing cells in the developing brain. Briefly summarized, these theories go something like this:[2]

The Immune System Theory. The brain and the spinal cord (which is composed of nerves that originate in the brain and travel all the way to the lower back) are encased in a sack called the meninges. This sack is made of a very tough but flexible material, a biological Saran Wrap. The primary job of the meninges is to protect the brain and spinal cord from invaders—bacteria, viruses, toxins, and so forth. It is the gatekeeper protecting the control center of our body.

There is another layer of protection around the brain and spinal cord, this one much more inconspicuous. It is a layer of cells surrounding the brain, held together by particularly tight junctions. This network of cells and cellular glue allows certain molecules into the brain but keeps most out. This is how it earned its nickname: the blood-brain barrier. The meninges are a solid, physical barrier while the blood-brain barrier is subtle and invisible. If the meninges are the gatekeeper, the blood-brain barrier is the key master.

Data in animal models has demonstrated that the radiation emitted from cell phones can cause damage to the blood-brain barrier. In 2002 the first study to look at human cells rather than rat cells (the prior standard) found that exposing these cells to one hour of mobile phone radiation caused changes within blood vessel walls, allowing tiny molecules to pass through the normally impermeable blood-brain barrier and into brain tissue. The authors concluded that repeated occurrences of these events—remember, it only took one hour of exposure to see changes—on a daily basis over years could lead to accumulation of brain-tissue damage and ultimately pose a true health hazard (Leszczynski 2002). If a cell phone's EMF breaks down the blood-brain barrier, one of the body's most important defense mechanisms is violated, and this may play a role in the formation of brain disease.

Another theory of how EMFs may affect the immune system involves the entire body and not just the brain. In this model, radiation results in an immune cascade, a kind of domino effect within the immune system. The radiation causes inflammation and allergic-type reactions, much like when you catch a cold—your body responds with headache, runny nose, cough, and congestion. The data seems to support acute inflammation: exposure to certain types of EMF does in fact stimulate an immune system response. But it is not clear whether this is short-lived, like a cold, or whether radiation causes chronic inflammation. Does exposure to EMF start a chain reaction within the immune system that continues on indefinitely? If it does, EMFs may be a trigger for chronic inflammation. Just as people with arthritis or lupus or inflammatory bowel disease are chronically inflamed, can people regularly exposed to certain types of radiation enter into this vicious cycle? No one knows for sure.

The Free Radical Theory. Most scientists agree that our cells are affected by certain electromagnetic fields through the action of free radicals. Free radicals are highly reactive atoms with unpaired electrons;

these electrons initiate chemical chain reactions that damage cells. The body seeks to pair up electrons so that there are as few free radicals as possible. This biological process is generally agreed upon. What experts do not agree upon is whether the body is capable of keeping up with—and repairing—that damage. If EMFs can delay the process, causing free radicals to stay free for longer, then this extra time gives them the potential to do more cellular damage. If the rate of damage from free radicals outpaces the rate of repair by the body, there will ultimately be an illness. So if EMFs are responsible for generating more and more free radicals, then they can be blamed for this imbalance. Free radicals have been accused of causing or contributing to dozens of diseases, including various cancers, atherosclerosis and heart disease, stroke and cerebrovascular disease, emphysema, diabetes, arthritis, osteoporosis, Crohn's disease, ulcers, aging, and the list goes on. But no one quite agrees about whether EMFs are a component of free radical–induced illness or how long it takes for this free radical damage to occur.

The Developing Brain Theory. You are born with all the brain cells (called "neurons") you will ever have. In fact, over time you lose neurons. These cells use electric signaling to communicate with one another. The loss of neurons over time allows the brain to hone its expertise, burn pathways of memory, and perfect skills. It is estimated that a newborn baby has about one hundred million neurons; by the time that person reaches eighty, she will have 5–10 percent fewer (only about ninety million to ninety-five million) neurons. You never get new neurons in the brain—you only lose them.

But there are other cell types in the brain that do divide, called "glial cells." Any cell that divides is more susceptible to cancer because cancer often comes from an error during cell division.

In a child's brain, the neurons are pruning themselves more rapidly and the glial cells are dividing more furiously than those in an adult brain. Therefore, both of these cell types are at risk for developing into cancers. In an adult brain, by contrast, neuron tumors are quite rare. The majority of adult brain tumors are glial cell tumors, because glial cell division outpaces neuron pruning. Other adult brain tumor types include metastatic tumors (tumors that started in other parts of the body and have spread to the brain) and meningiomas (tumors of the meninges, the sheath covering the brain and spinal cord).

Many scientists believe that EMFs affect a child's developing brain more readily than a mature adult brain. This assumption is based on two facts. First, a child's skull is thinner than an adult's, so radiation has less of a barrier to cross. Second, the constant changes in a child's brain evidenced by leaps in development mean that the cells in the brain are more susceptible to injury and therefore more likely to become cancerous. I actually don't agree with this second statement. As we master a skill, neurons grow in length, with their long arms reaching out to one another and making connections. This should have nothing to do with the chromosomes in the neurons, the blueprints locked inside the cell's center. Cancer occurs when the chromosomes—not the arms—of the cells mutate. But it's still a theory floating out there, so I add it to be comprehensive.

Each of these theories—the immune system theory, the free radical theory, and the theory about developing and growing brains—offers a plausible explanation, but none has been proven. Part of the challenge in finding a cause is that the majority of brain tumors take a long time, often years, to form. Frankly, the theories about EMFs causing cancer are superhypothetical, since the studies don't cover enough time and there's no proven link. All we know for sure is that EMFs transmitted by cell phones can affect cells.

There also needs to be more data collected about all the countless things in our world that can turn cells bad: exposure to other sources of EMF, including the use of WiFi and other wireless devices, personal lifestyle, types of food ingested, exposures to plastics and other toxins, and so on. Only then can experts really begin to make cause-effect pairings. And to prove any of this will take many more years of good data collection—studies that span fifteen, twenty, thirty years. Right now many scientists are looking specifically at how long you talk on your cell phone and how you hold it. There's probably going to be a whole lot more to it than that.

It may turn out that one of the three theories is more correct than another or that the truth comes from a combination. These aren't the only explanations out there either. There are other theories circulating offering alternate mechanisms for EMF-induced damage. What seems certain is that *something* will pan out. Based upon everything I have read, it is impossible that EMFs will turn out to have no biological impact.

That said, however, biological impact doesn't mean danger. If EMFs are proven to cause cell damage, it doesn't automatically mean that cell phones cause cancer. This last point is critical: we are eager to find ways to make our world a safer place, but it does no good to jump to conclusions and change our behavior if it will make no difference to our health and well-being in the long run. Why demonize the cell phone if we aren't sure whether it causes disease? What if brain cancer has nothing to do with cell phones and is actually a result of something else? This is why the debate goes on.

Despite the arguments over whether cell phones are dangerous, there are regulations in place to protect public safety. As explained at the start of this chapter, all electric and wireless devices are regulated because of the known dangers of excessive heat and induced electric currents. The federal rules reflect these two agreed-upon risks. No one wants to loosen these regulations; the back-and-forth is over whether to tighten them by adding nonthermal effects and lower-frequency EMFs to the list of safety criteria. Until recently, these variables had been considered benign (or hadn't been considered at all). Today in the United States there is little agreement among scientists, epidemiologists, and environmentalists as to whether nonthermal effects and low-frequency EMFs represent real dangers. Current regulations are only going to change if these risks are proven to be real and if specific agencies agree.

The agencies at play are the FCC and the **FDA**. The FCC has dictated much of the U.S. cell phone safety policy. The FCC has mandated that cell phones and other wireless devices meet EMF-exposure guidelines designed to protect against thermal injury. Whereas high levels (>1,000 mG) of exposure to EMFs can heat tissue so much that it essentially cooks it, low-level exposure does not. Basically, avoiding heat-related injury is the predominant criterion for the FCC rules, and because this is their yardstick, their current maximal EMF level is 904 mG. Understanding the precise units is not very important for most people—just keep in mind that number: 904.

Wireless phones emit EMFs while being used. They also emit a minimal amount of EMF when in the standby mode, though the intensity is much less than when you are on a call. The FDA has generally agreed with the EMF limits set by the FCC. But unlike the FCC, the FDA recognizes the potential nonthermal effects of EMFs and the concerns about low-level EMF exposure. It acknowledges the evidence but doesn't see a reason to change the current safety rules. Why not? Because according to the FDA, there isn't enough data proving that low-level EMF exposure causes health effects. Remember, scientific theories are only considered proven when there have been multiple studies showing the same results.

The FDA doesn't have a lot of motivation to change the rules either. By law, the FDA is *not* required to review the safety of wireless-radiation-emitting consumer products before they can be sold. This is totally different from the FDA's role with drugs and medical devices—those products have to be studied and restudied before they go on the market. When it comes to wireless technology, the FDA only has to take action if phones or other devices are shown to emit EMFs at a level that is proven hazardous to the user. The agency hasn't done this, claiming that the existing data doesn't justify it. It seems to be a bit of a catch-22: the FDA doesn't have to study the effects of EMFs emitted by wireless devices before they go on the market, so there is limited safety data; and because there is limited safety data, the FDA doesn't see a reason to change its level of oversight.

The one thing I can say for the FDA is that it has placed the ball in the court of the wireless industry and strongly encouraged companies to do something with it. The FDA has urged manufacturers to support research into possible biological effects of EMFs emitted by wireless phones, to design wireless phones to minimize users' EMF exposure, and to inform users of wireless phones about possible effects on human health. So the FDA is willing to take a much broader view of cell phone safety than the FCC, even though it won't change its policy.

But around the world, the situation is quite different. Over the past decade, **WHO** has become increasingly influential in electromagnetic-radiation policy. In June 2001, the International Agency for Research on Cancer (**IARC**), a branch of WHO, announced that extremely low-frequency magnetic fields could cause cancer in humans.[3] In July 2007, WHO took this to another level, releasing a policy advisory emphasizing the importance of EMF exposure limits because of established acute effects (and likely chronic effects) of the magnetic fields generated by extremely low-frequency EMFs.

So WHO, a leading global health organization, is unafraid to point a finger at electromagnetic radiation, particularly in the wake of sky-rocketing access to cell phones. Unlike the FCC and the FDA, WHO has become vocal about the possible health effects of EMFs. The organization implies that current rules in the United States only address public concern but don't sufficiently address public safety. Still, the shift at WHO hasn't had much of an impact on U.S. policy.

The rest of the world, though, has started to listen. In 2007, about the same time that WHO made its statement, came the publication of *The Bioinitiative Report*. The report evolved from the Bioinitiative Working Group, a collaboration of fourteen scientists, public health officials, and consultants from Europe, the Americas, and Asia. The group set out to answer a basic question: why, if there are thirty years' worth of scientific studies and good resulting data about the increasing danger of EMFs, have there been no significant changes in safety standards? The group's goal was to educate countries around the world in an effort to change outdated safety standards that do little to protect human health.

The Bioinitiative Report had a dramatic impact in Europe: within weeks of its publication, governmental agencies were recommending new safety standards for electric power and wireless devices. In multiple countries, organizations have changed safety limits for all sorts of electric and wireless devices, factoring in likely dangers of nonthermal radiation and low-frequency EMFs. Individually, countries have established working groups that advise evolving governmental policies.

In the United States, though, the report went largely unnoticed, and no sweeping reforms were recommended. Remember that number, 904? This is the currently recognized upper limit of mGs deemed safe by the FCC. *The Bioinitiative Report* invokes more than two thousand studies published in scientific journals to suggest over and over that the actual upper limit of safety should be 1 mG, and many think that's too high. This represents a nearly thousandfold difference from the current standard. At this point, many industrialized nations have taken notice and amended their EMF standards for devices, including cell phones. It is time for the United States to follow suit.

WHAT IS THE BOTTOM LINE?

Nobody knows for sure whether cell phones and other sources of EMF cause cancer. It is difficult to know what to do when there are two completely different schools of thought on the topic—one that says that because EMF affects cells we need to change our behavior and another that doesn't want change until there is more research. I don't think the data conflicts, though: everyone agrees that the EMF generated by a cell phone can impact the body. Their disagreement is over what to do about it.

Just because cells in a laboratory respond to EMF doesn't mean that exposure to EMF makes people sick. There are lots of studies showing no link between EMFs and cancer. But here's the problem: these are short-term studies, many with fewer than three years of follow-up. Newer data collected over longer spans of time (around ten years) shows an association with cancer. The longer the studies, presumably the more damning evidence we will see. But we cannot just change policy every time a new study is published. And it takes time to replicate studies.

This whole story smells very familiar. Remember when smoking was considered safe? There were clear correlations between smoking and all sorts of diseases—lung cancer and asthma, to name just two—before it was technically "proven" to be a cause. Many say that the government waited far too long to campaign against smoking and that thousands of lives would have been saved if something had been done before the evidence was rock solid.

Cell phones could be the cigarettes of the new millennium. There are already good theories postulated as to how they inflict their damage, and public interest is growing. Life without cell phones seems as unimaginable as, well, life without cigarettes seemed in 1950.

Cell phones won't go away, but you do need to learn to use them more safely. If you must conduct extended conversations by wireless phone every day, place more distance between your body and the source of the EMF, since the intensity of the field drops off dramatically with distance. Use a wired earpiece and carry the phone away from your body or inside a holster to contain the radiation. At home, move back to corded phones. People forget that cell phones aren't dangerous just because they have the word "cell" in the title—they are presumably dangerous because you are placing your brain into the electromagnetic field generated by an antenna. The same risk exists with your home cordless phone.

For parents, here's yet another reason to keep your kids away from a cell phone. Young brains may be more susceptible to cellular damage—and perhaps cancer—than older brains. It is unclear why (or even if) this is the case. Maybe it is because your child's skull is thinner than yours, making it easier for EMFs to penetrate. Maybe it has something to do with the dividing and pruning of the brain cells. Or maybe we just have to consider that our kids will have a lifetime of cell phone use, which means more exposure to EMFs.

Cell phones scare me—of everything in this book they are what I consider the most truly dangerous. But I'm not just thinking in the future tense—there are immediate dangers associated with cell phones too. While it is possible that cell phone use will be linked with cancer, it is certain that cell phone use causes car accidents. Every year in the United States there are approximately 330,000 automobile accidents that result directly from cell phone use. Of these, 2,600 wind up in fatalities. The cancer debate may be frightening, but the use of cell phones in any way, shape, or form while driving is downright dangerous.

The following was posted on the University of Pittsburgh Medical Center's Web site in the summer of 2008. Dr. Herberman was the first U.S. cancer center director to release an advisory like this. The memo received much attention throughout the medical community, but not many other cancer center directors followed suit.

Important Precautionary Advice Regarding Cell Phone Use

FROM: RONALD B. HERBERMAN, MD

Recently I have become aware of the growing body of literature linking long-term cell phone use to possible adverse health effects including cancer. Although the evidence is still controversial, I am convinced that there are sufficient data to warrant issuing an advisory to share some precautionary advice on cell phone use.

An international expert panel of pathologists, oncologists, and public health specialists recently declared that electromagnetic fields emitted by cell phones should be considered a potential human health risk. To date, a number of countries including

France, Germany, and India have issued recommendations that exposure to electromagnetic fields should be limited. In addition, Toronto's Department of Public Health is advising teenagers and young children to limit their use of cell phones, to avoid potential health risks.

More definitive data that cover the health effects from prolonged cell phone use have been compiled by the World Health Organization, International Agency for Research on Cancer. However, publication has been delayed for two years. In anticipation of release of the WHO report, the attached prudent and simple precautions, intended to promote precautionary efforts to reduce exposures to cell phone electromagnetic radiation, have been reviewed by UPCI experts in neuro-oncology, epidemiology, neurosurgery and the Center for Environmental Oncology.

Practical Advice to Limit Exposure to Electromagnetic Radiation Emitted from Cell Phones

• Do not allow children to use a cell phone, except for emergencies. The developing organs of a fetus or child are the most likely to be sensitive to any possible effects of exposure to electromagnetic fields.

• While communicating using your cell phone, try to keep the cell phone away from the body as much as possible. The amplitude of the electromagnetic field is one-fourth the strength at a distance of two inches and fifty times lower at three feet. Whenever possible, use the speakerphone mode or a wireless Bluetooth headset, which has less than one one-hundredth of the electromagnetic emission of a normal cell phone. Use of a hands-free earpiece attachment may also reduce exposures.

• Avoid using your cell phone in places, like a bus, where you can passively expose others to your phone's electromagnetic fields.

• Avoid carrying your cell phone on your body at all times. Do not keep it near your body at night such as under the pillow or on a bedside table, particularly if pregnant. You can also put it on "flight" or "off-line" mode, which stops electromagnetic emissions.

• If you must carry your cell phone on you, make sure that the keypad is positioned toward your body and the back is positioned toward the outside so that the transmitted electromagnetic fields move away from you rather than through you.

- Only use your cell phone to establish contact or for conversations lasting a few minutes, as the biological effects are directly related to the duration of exposure.
- For longer conversations, use a land line with a corded phone, not a cordless phone, which uses electromagnetic-emitting technology similar to that of cell phones.
- Switch sides regularly while communicating on your cell phone to spread out your exposure. Before putting your cell phone to the ear, wait until your correspondent has picked up. This limits the power of the electromagnetic field emitted near your ear and the duration of your exposure.
- Avoid using your cell phone when the signal is weak or when moving at high speed, such as in a car or train, as this automatically increases power to a maximum as the phone repeatedly attempts to connect to a new relay antenna.
- When possible, communicate via text messaging rather than making a call, limiting the duration of exposure and the proximity to the body.
- Choose a device with the lowest SAR possible (SAR = "specific absorption rate," which is a measure of the strength of the magnetic field absorbed by the body). SAR ratings of contemporary phones by different manufacturers are available by searching for "sar ratings cell phones" on the Internet.

WHAT'S IN MY HOUSE?

After researching this chapter, I made some changes in my everyday life. I do my best to use a wired earpiece every time I talk on the cell phone. My husband knows his marching orders, so he plugs in the earpiece as well.

While I was writing, California passed a hands-free cell phone law. Connecticut, New Jersey, New York, Washington, the District of Columbia, and the Virgin Islands all have the same rule. Drivers must have two hands on the wheel and that means drivers cannot hold their cell phone to their ear. In my car, I have a surround-sound Bluetooth that runs through my stereo. There probably will not be a significant drop in car accidents related to this new law because the problems with cell phone use in the car have to do with lack of concentration while driving and talking, not with how you hold the phone. Chatting on a cell phone is distracting. Dialing or texting is even more so.

As for my home phones, we are gradually switching over to corded phone sets. We still have a cordless phone in the kitchen, but I do my best to not use it.

My kids are too young to ask for cell phones, but I know it's just a matter of time. When they do, there will be some rules. First, they can never use one without a wired earpiece (I don't even let them talk on mine). Second, I am going to preprogram a few numbers—like my number, their dad's, 911, and so on—but my kids won't be able to dial anything else. This type of parental control significantly decreases use. It is a great way for parents to balance the safety benefit of a cell (you can reach your child at any time) with the possible dangers (you don't want your child talking on it nonstop).

Chapter 13

Flame Retardants

WHAT IS THE QUESTION?

When flame-retardant chemicals were first marketed, they were considered a giant leap in the world of safety. Fires could be delayed by coating clothing, electronics, mattresses, and dozens of household items with retardants. Although the items could still catch fire, it would take much longer. This extra time would allow people to smell smoke and evacuate a burning house or building before everything went up in flames.

There is no doubt that flame retardants have saved thousands of lives. But like everything else in our world these days, these chemicals have fallen prey to intense scrutiny. This is not unreasonable: flame retardants are literally everywhere, and if they pose a health risk, consumers need to know.

So are flame-retardant chemicals more dangerous than they are safe? What exactly are the concerns? Some say these chemicals are partially responsible for the dramatic rise in autism. Are the rumors of brain toxicity true?

WHAT IS THE DATA?

Flame-retardant chemicals are made from polybrominated diphenyl ethers (often called by their acronym **PBDEs**). Within this family of ethers, three specific flame retardants are used most frequently: pentaBDE, octaBDE, and decaBDE. DecaBDE (often called simply "deca") is by far the most predominant type, accounting for more than 80 percent of the world's PBDEs. It is generally used for electronics and electronic enclosures like wire insulation. OctaBDE is found in plastics and small appliances, while pentaBDE is in foam cushions for upholstery.

The only distinction among these family members is that each has a different number of bromine atoms per PBDE molecule—penta has five, octa eight, and deca ten. Otherwise, they all share the same function: to slow the spread of fire.

When materials have flame-retardant coatings, it takes longer for them to catch fire. When there is a house fire and the furniture and clothing and electric equipment are made with PBDEs, the family has more time to escape because fire in one part of the house takes longer to spread to another.

PBDEs were first commercially produced in the 1970s. Almost immediately they appeared in dozens of household items: foam padding, plastics, appliances, fabrics, upholstered furniture, food packaging, computer keyboards, toys, and clothing, to name a few. But over time these items age—furniture becomes tattered, keyboards lose keys, toys break apart, and so on. With wear and tear, PBDEs become part of the dust around the house.

This poses a potential problem with a chemical that was once considered a panacea: PBDEs in dust are inhaled—as house dust often is—and absorbed into our bodies. Children actually inhale house dust much more readily than adults. Why? Babies crawl across the floor, breathing in the dust and debris that have settled on it. Older kids walk but constantly put their dirty hands in their mouths, licking particulate chemicals off their fingers. Studies document that per unit of body weight, children carry more PBDEs in their bloodstream than adults. In 2008 the Environmental Working Group published a study showing that American children between one and four years of age bear the heaviest burden of flame-retardant pollution in the industrialized world, with PBDE levels three times higher than those of their parents.

Another source of PBDE exposure is through our food supply, particularly in meat, fish, and dairy products. PBDEs enter the food chain through environmental pollution and accumulate in fat tissue. A general rule is that the fattier the meat, the more likely it contains PBDEs.

Food PBDEs affect adults and kids fairly equally. One might think that young babies are spared this exposure because they aren't yet eating solids, but infants get plenty of PBDEs through breast milk that is produced from the fatty breast tissue. Over the past few years, in fact, breast milk has been found to contain increasing concentrations of PBDEs. A 2003 report stated that among American women PBDE levels in breast milk are ten to one hundred times higher than the average levels measured in European women (Schecter 2003).

The benefits of PBDEs outside our bodies were clear from the start, but no one really knew the risks of PBDEs to our insides. Now that scientists know that most of us breathe, eat, and absorb these chemicals, researchers have begun to study potential biological effects. But studies take time. To estimate what the biological effects of PBDEs might be, researchers started by reanalyzing data collected on similar, though not identical, chemicals. This is why when you read about PBDEs you come across a lot of information about their close cousins, **PCBs** (short for polychlorinated biphenyls).

PBDEs and PCBs share a very similar structure, which is precisely why researchers used PCBs as their model. Often—but not always—chemicals with similar structures affect the body in nearly the same way. There is a lot of data about PCBs, most of it damning. PCBs can cause acne or irritation of the skin, affect liver function, and alter the immune system's ability to respond to infection. They have also been associated with nonspecific symptoms such as headache, fatigue, and cough. Some babies exposed to high levels of PCBs in the womb or following breast-feeding have slower than average gross motor development. And there is evidence that PCBs act as **endocrine disruptors**—chemicals that may affect the way hormones work in the body—causing reproductive problems and even some cancers.

A momentary aside about endocrine disruptors: The term "endocrine disruptor" is very vague and can mean many different things. In some cases, endocrine disruptors actually mimic a hormone (like estrogen), but in other cases they simply affect the way some hormones work in some people's bodies. Endocrine disruptors appear in half a dozen chapters in this book, partly because they are all around us and partly because they provide a new framework for looking at how chemicals affect living things. Each member of this club works differently, affects tissues in its own way, and is linked with its own set of possible diseases and cancers. Endocrine disruptors include **phytoestrogens** in soy, **phthalates** and **parabens** in cosmetics, **bisphenol A** in plastics, and now PCBs and PBDEs.

Back to our story. The association between PCBs and PBDEs has put a cloud of suspicion over flame retardants. The EPA, National Cancer Institute, and others have published statements that PCBs are known or potential carcinogens. As a result of statements like these, PCB production was banned in the 1970s, and the criticism of this chemical continues more than thirty years later. So wherever there are clear similarities between PCBs and PBDEs, there is heightened (and reasonable) concern.

One way PCBs and PBDEs are notably similar is that both bioaccumulate. This is another way of saying that they stick around in the body. The significance of bioaccumulation is huge: any chemical that remains in the body has greater potential to result in illness than a chemical that leaves the body. Add to this the fact that over time even low-dose exposure to a chemical that bioaccumulates will result in gradually increasing levels of the chemical in your body. If the body isn't capable of getting rid of a chemical and it slowly builds up over years and years, its potential health repercussions continually increase. This seems to be the case with PBDEs— now we just have to figure out what those health consequences are.

Associating PCBs and PBDEs is not without rationale: as I said before, the two chemicals look very similar. But as with other remarkably similar seeming chemicals discussed in this book (like estrogen and phytoestrogen, for instance), just because they look alike doesn't mean they act alike. Scientists turned to PCBs as a point of comparison, but PBDEs are *not* PCBs and they are entitled to their own scrutiny.

Unfortunately, there are very few completed studies that look at the human health effects of PBDEs. The research is still under way. But there are many animal studies suggesting that PBDEs act as endocrine disruptors. In these studies, lab animals exposed to high doses of PBDEs have exhibited thyroid hormone effects, reproductive effects like decreased sperm function and ovarian abnormalities, and neurological effects ranging from hyperactivity to impaired brain development and learning difficulties. Early data suggests that the human studies will confirm PBDEs' place in the family of endocrine disruptors. The studies also suggest some nonendocrine toxicity as well, such as skeletal malformations and liver tumors.

Much has been made of the potential health risks of PBDEs. These chemicals are ubiquitous—in every corner of our homes, offices, schools, restaurants, shops, and cars. If there is a health hazard, consumers deserve to know. Parents in particular are eager—no, anxious—to figure out whether PBDEs are to blame for any number of health issues facing their children, from SIDS to the early onset of puberty to the dramatic explosion in autism.

On the subject of autism in particular the CDC has said that infant animals exposed to PBDEs may demonstrate subtle behavior changes. But they caution that it is impossible at this stage to tell how that might translate for humans. PBDEs could cause effects on the central nervous system in humans like delayed development or even autism. Or perhaps these chemicals affect the thyroid gland, which in turn can delay development. Or PBDEs could do both, because the brain and the thyroid are quite intertwined, with maturation of the nervous system depending partly on thyroid hormone. There are many people involved in the war on autism—parents, physicians, scientists, and journalists—who have suggested a link with PBDEs. It is a reasonable theory, but there is no data yet conclusively proving that PBDEs cause autism.

The majority of human data on PBDEs is preliminary data; most completed studies are on lab animals. It is difficult to translate the data from animals to humans, not just because animals are physiologically different from humans, but also because the doses used in animal studies are often much higher than those found in the real world. PBDE dosing isn't exactly standardized either. In different parts of the globe, PBDE exposure varies widely. Recent studies have shown that PBDE levels in Americans are higher than in people from most other countries—on the order of ten to one hundred times higher than in Europeans—and continue to rise.

So what do we do while we are waiting for the data? It seems inappropriate to prematurely vilify PBDEs, but if they really are potent endocrine disruptors—not to mention contributors to other diseases—isn't it best to avoid them? The short answer is yes: avoiding something with unknown safety data is almost always the prudent thing to do. But realistically, this is impossible. PBDEs are everywhere.

One way to begin to reduce exposure to this family of chemicals that will likely turn out to have health implications is to decrease its production. This has already begun. Over the past few years, penta- and octaBDE have been all but phased out of production. By 2004 they were no longer manufactured in Europe and industry voluntarily stopped production in the United States.

DecaBDE, on the other hand, has not faced this fate. This is because deca, with its extra bromine molecules, is bulkier than penta- or octaBDE. Scientists believed that this bulkiness explained why deca seemed to be absorbed at a much slower rate by the human body. And for this reason, the PBDEs were stratified—manufacturers of flame retardants presumed that decaBDE was safer than penta- or octaBDE.

DecaBDE is the last remaining commercially produced flame retardant in the polybrominated ether family. But as data continues to emerge, even deca may soon go out of production. While it is still in worldwide commercial use as a flame retardant, it was banned in Europe in 2008.[1] Some states in the United States have movements afoot to ban decaBDE as well. Its absorption by the body is slower than the other PBDEs, but some of the chemical is still absorbed. The health and safety implications remain unclear.[2]

We have almost forty years' worth of PBDE-coated mattresses, carpets, clothes, wires, and keyboards filling our homes, our offices, and our schools. Future exposures may decrease as octa- and pentaBDE are no longer used, but past production continues to affect us. We live with this chemical all day, every day.

The story goes beyond PBDEs, way beyond. We live surrounded by dozens of other chemicals. According to a 2006 article in the *Lancet*, chemical pollution—not just PBDEs—is a serious threat to the human brain (Grandjean 2006). There are 202 industrial chemicals identified so far with the capacity to damage the human brain, and many believe that the toxic effects of these chemicals on children have been overlooked. Teasing each and every one of these chemicals (and their effects) apart is a gargantuan task.

PBDE is one example of an industrial chemical that was designed to save lives, not harm them. Though there may be evidence that PBDEs carry some toxicity—particularly neurotoxicity to the developing brain—we need to weigh their risks against their benefits. Fire can destroy, maim, scar, and kill. There is no option that involves *not* using a flame retardant in our homes, our clothes, and our electronics. There are newer chemical flame retardants available—specifically ones without polybrominated ethers. But just because something new is free of the suspicious ingredient doesn't make it safe. With these latest chemicals, there is often *no* data. That means there is no negative data, but there is no safety data either.

WHAT IS THE BOTTOM LINE?

In the United States, more than one million people get burn injuries each year, and burns are the fourth leading cause of death from unintentional injury. In 1979 nearly six thousand people died in fires; in 1999, just twenty years later, the number had been reduced by nearly half (Eisenberg 2002). Education, fire alarms, and basic safety training all deserve a great deal of credit. But flame-retardant chemicals also warrant a share of the praise. Flame retardants slow the progression of fire so people can get out of harm's way in time.

The most common flame retardant in use—the family of PBDEs—may turn out to have health consequences. Researchers are furiously collecting data on humans. Animal studies suggest that PBDEs are not only absorbed but act as endocrine disruptors with the potential to affect our hormones and our reproductive organs. They may also directly affect the brain and, in children, the development of the brain. Could PBDEs be one contributor to the explosion of autism? We simply do not know.

What we do know is that we are exposed to PBDEs mostly through house dust and food. House dust is the end point for many chemicals in our homes. If you stop and think about what you are inhaling, you may be more motivated to turn on the vacuum or mop the floor. Realistically, you cannot get rid of all the PBDE-coated objects in your home. But you can be more aggressive about cleaning, which should minimize your exposure.

You can also get rid of old toys, cushions, electronics, and plastics. If something looks like it is starting to decompose, then toss it. If you are eager to replace something in your home in order to make a dent in the problem, the first thing I would recommend is to trade in your mattresses for some made with **organic** cotton. Getting rid of foam bedding and PBDE-treated mattresses minimizes your child's (and your own) exposure for ten to twelve hours every day. But there are lots of varying opinions about the quality of organic mattresses, and many people argue that just because the mattress says "organic" doesn't mean that it's free of chemicals. So this purchase may require a little legwork on your part.

To further minimize contact with PBDEs, try to keep your child away from electronics as long as possible. That means that if you are a new parent, you shouldn't e-mail with the baby on your lap: you may feel accomplished because you are getting work done while bonding with your child, but if your infant's face is inches from the keyboard, he is inhaling PBDE debris.

Likewise, enforce hand washing aggressively. A toddler or school-aged child who comes home and eats a snack with her hands is ingesting more than just dirt because her hands are also layered with PBDE residue from the day. Teach the importance of hand washing and you will reduce your child's ingestion of PBDEs and many other chemicals.

Our other primary source of PBDEs is in our food supply. Meat and fish—particularly fatty varieties—accumulate large amounts of PBDEs from the environment. Pick leaner cuts, trim the fat off the perimeter, and when you cook use broilers that help drain fat. All of these strategies will reduce your family's exposure.

PentaBDE and octaBDE are sufficiently worrisome that they are no longer in production in many parts of the world; decaBDE is still produced and used widely, but the European Union banned it in 2008, and several states in the United States have movements afoot to do the same. Remember, though, data about their effects on humans is not widely available yet. While it is fair to suspect that PBDEs may have health implications, no one agrees as to what those implications are.

Before you throw out everything in your house that could possibly have PBDEs, remember that slowing the spread of fire saves many more lives than PBDEs harm, at least as far as we know right now. Choosing items with no flame retardant at all is actually the least safe option.

Critics often jump to the next best thing—in the case of flame retardants they advocate just about any variation without PBDEs. There are lots of "non-PBDE retardants" available on the market. Even though it is logical to try to avoid a chemical that may have associated health risks, it doesn't mean that the alternative is any safer. A non-PBDE-containing flame retardant may very well be on the receiving end of bad press in another few years.

WHAT'S IN MY HOUSE?

My house is chock-full of PBDEs: they are in our electronics, computers, bedding, kitchen appliances, and on and on. This doesn't thrill me. But I recognize the importance of flame retardants, and I am not yet sufficiently convinced by the data that alternative chemicals are any safer. So my goal is to minimize my family's exposure to all of these chemicals. Friends mock us because we are the cleanest family they know. But the added benefit to vacuuming the dust bunnies and enforcing hand washing is lower PBDE exposure.

Chapter 14

Lead

In June 2007, 1.5 million Thomas & Friends toys were recalled because of leaded paint. Many of the red-painted engines and signposts (and some yellow ones too) that were manufactured in China between 2005 and 2007 included a surface paint that contained lead. If enough lead is ingested, it can have toxic effects. The paint posed a risk because young children who play with the Thomas toys are likely to put them in their mouths, biting or licking or chewing the leaded paint. Paint also ages, and the chips can form a fine lead dust that's easy for children to inhale.

The recall caused uproar and was the star health hazard of the moment. But the uproar was short-lived and, in hindsight, the attention on the Thomas toys was somewhat random. The Thomas recall was just one of thirty-one published by the U.S. Consumer Product Safety Commission (**CPSC**) in the same month. The CPSC Web site lists more than two hundred lead-related recalls since 2001, the vast majority in 2007 and 2008. In 2007 alone, more than 9 million toys were recalled because of lead, including 1.5 million Thomas units and 1.5 million Fisher-Price toys (among them Dora the Explorer and *Sesame Street* characters Big Bird and Elmo). Many other items on the list were much less well publicized, including pajamas, jewelry, key chains, art sets, figurines, helmets, and puzzles. Thomas was just the tip of the iceberg.[1]

The lead debate peaked and died quickly. Just two months later, the public turned its attention to the dangers of **bisphenol A** in baby bottles and other plastics. Within weeks of the initial Thomas commotion, lead was already forgotten by the general public. But the Thomas recall inspired a tremendous amount of renewed (albeit short-lived) concern about heavy-metal poisonings. Lead—at least in this country—was considered a thing of the past. The days of leaded gasoline and leaded paints were long gone. The worry that the Thomas recall generated was legitimate because lead has been shown to cause neurological disease, learning and attention issues, and loss of IQ points in children. But did the Thomas recall warrant such distress? You and I grew up in a highly leaded world. Did the exposure to lead in the Thomas engines (and other lead-tainted toys) represent even a fraction of what we were exposed to as children? Is a limited exposure to lead really something dangerous?

WHAT IS THE DATA?

Lead is a heavy metal that accumulates in the human body. It is toxic to the nervous system, especially among young children. High blood-lead levels are known to cause cognitive deficits (with loss of IQ points), attention and behavior issues, and gastrointestinal complaints like stomach upset, nausea, and diarrhea. In adults lead is associated with blood clotting disorders, high blood pressure, and damage to the reproductive organs. Some studies suggest that excessive lead is associated with psychiatric diseases like schizophrenia and with delay of puberty in girls.

Lead exposure comes primarily from paint, dust, soil, and drinking water. In the past, gasoline and leaded food cans were also significant sources.

In 1978 lead was taken out of all household paints. In the thirty years since this mandate, blood-lead levels have dropped by more than 82 percent, and the change in the composition of paint is credited with much of this decline. Why did this one intervention make such a big difference? Because household paint chips over time, and children often play with or pick at the chipped paint. The lead is transferred to their skin and then licked off their hands. Chipped paint also generates a significant amount of dust, especially around windows and doors. Toddlers who balance on these dusty frames—particularly the lower ledge of the windowsill, which is just about at toddler nose level—inhale the lead particles in the dust. Finally, houses' exterior paint gets mixed into surrounding soil. This happens as a result of chipping, dust formation, and general weathering. If there is leaded paint on the exterior of your house, there is often lead in the soil of the yard where your kid plays. Homes built before 1978 that have not been remodeled or completely stripped and repainted often still have layers of leaded paint. Therefore, while the availability of lead-free paint over the past thirty years has made an enormous difference in terms of blood-lead levels, paint-related exposure is still not at zero.

Tap water is another significant lead source. Like paint, pipes were also once lead based. Lead was used in the manufacture of piping and in the solder holding pipes together. The EPA banned lead in pipes and solder starting in 1986 but, again, unless a home has had a major remodel including replacing all plumbing any home built before the mid-1980s may still have leaded pipes.

Gasoline was another major source of lead contamination, and removing lead from gasoline no doubt contributed to the 82 percent drop in average blood-lead levels. Gas was leaded beginning in the 1920s. The first lead-reduction standards for gasoline were imposed in 1973, five years before lead-free paint was required. The gas policy, however, was a gradual one, with steady declines in lead content spread out over twenty years. As gasoline gradually became unleaded, so did car emissions. Exhaust provided a significant amount of aerosolized lead, which typically fell to the ground near roadsides and settled into the soil. The Clean Air Act of 1996 marked the final nail in leaded gasoline's coffin, banning the last leaded gasoline sold for on-road vehicles. Much roadside soil however, especially along highways, still contains a significant amount of lead.

Like gasoline, leaded food cans were once the standard. Starting in 1991, lead solder was no longer used to can goods, removing this source of contamination. Before 1991, though, lead leached into the food when the can was sealed with leaded solder. According to the **FDA**, between 1982 and 1996 daily intake of lead from food dropped 96 percent in two- to five-year-olds, from an average of 30 *ug* a day down to 1.3 *ug*.[2] The drop in adults was almost as high: a 93 percent reduction from 38 ug per day to 2.5 ug.

It would be ideal if we had zero lead in our bodies. But we don't. Old paint, soil residue, solder, and pipes continue to expose us to small amounts every day. How much is too much? The **CDC** has defined the maximal acceptable daily lead intake as 6 ug for kids under six years old, 25 ug for pregnant women (lead crosses the placenta and can be absorbed by the fetus), and 75 ug for adults. The reason adults can tolerate a higher daily exposure is that only about 10 percent of the lead reaching their digestive tract is actually absorbed. By contrast, kids absorb anywhere from 30 to 75 percent. And children are getting much more lead into their digestive tracts in the first place, as they tend to put dirty fingers and other objects in their mouths. When a child inhales leaded dust, 50 percent of that lead is absorbed; but when he plays with leaded paint chips, less than 1 percent goes through the skin of his hands.

These numbers don't mean much to you and me because lead quantity does not appear on labels or in other disclaimers. We have no idea what our daily exposure really is. The best way for people to monitor their lead exposure is to measure their blood-lead level. This number is reported in micrograms per deciliter (ug/dl). A microgram is one-millionth of a gram; a deciliter is equal to about 3.5 ounces of blood (a little less than half a cup). In 1991 CDC recommended the current maximum acceptable lead level. In adults lead should be less than or equal to 25 ug/dl; in children less than 10 ug/dl. In 2005, the AAP policy statement on lead exposure reconfirmed this threshold for children. High levels require intervention: for a child with a lead level higher than 10 ug/dL there needs to be education about exposure, close follow-up, and retesting; if it is higher than 15–20 ug/dL medical evaluation and lead abatement in the home; for even higher levels medical intervention is necessary.

In 1993 the **AAP** endorsed the CDC recommendation and did something about it, recommending routine lead screening for all nine- to twelve-month-old children. The goal was to identify the children with elevated lead levels in order to locate the sources of this well-known toxin so that parents could remove the sources and, ultimately, reduce potential harm.

Over the past fifteen years, there has been a call to reduce the acceptable blood-lead level in children. It has become increasingly clear that a lead level as low as 2 ug/dl represents unnecessary exposure to the heavy metal. Therefore, a recent trend (though not AAP policy) has been to screen all children for lead exposure and to consider recommending interventions when the level comes back at or above 2 ug/dl.

It stands to reason that if paint, cans, piping, and gas no longer contain lead, at some point lead exposure should drop close to zero. It makes sense that lead continues to cause problems in old houses, but as those houses are renovated or rebuilt the exposure should decrease until it is gone, right? Wrong.

Until August 2009, paint was considered "safe" for children's products if it had less than 600 **ppm** lead (which equals 0.06 percent lead by weight). This is not zero. A recall—like the Thomas recall in 2007—was generated when the lead content of paint exceeded 600 ppm, but anything under that was essentially called "lead free." After August 2009, the maximum acceptable lead concentration in paint fell to 90 ppm.[3] While this is an improvement because it gets us closer to zero (and everyone seems to agree that lead is bad), the industry standard for "lead-free" paint is not exactly free of lead.

It is nearly impossible to find a written statement of how much lead was in the Thomas paint, but it must have been higher than 600 ppm. And if a child ingested enough paint by chewing on those toys, there was certainly a risk of a slightly elevated blood-lead level. To prevent unnecessary exposure to a known toxin, it is reasonable to create strict standards. That said, you probably don't need to worry about absolutely every Thomas and Dora and *Sesame Street* toy in your house that could have possibly been on the recall lists. We're talking about a pretty low level here. And with the new laws in place, exposures will be even lower in the future.

Your child's primary exposures are residential lead in paint, solder, and soil. If your child is tested for lead and found to have an elevated level, it is unlikely that a Thomas & Friends toy was the primary cause. Yes, we should be limiting imports tainted with lead and be notified when these toys get through. But recalled toys are not major sources of lead poisoning in this country. We are, in fact, doing a good job of revising our own standards about what constitutes "safe" and "lead free" on our own shores, too. The goal, after all, is a blood-lead level as close to zero as possible.

WHAT IS THE BOTTOM LINE?

If—and when—there is another toy recall for leaded paint, don't panic. Clear the toys out of your house if you have a toddler who puts everything in his mouth. But if you have older kids who no longer chew their toys and who wash their hands well, or if you only have a few potential hazards here and there, you don't need to break out in a cold sweat.

It is a better use of your time and energy to consider more likely sources of lead. There are lots of Web sites with good basic information about identifying lead risks in your home. If you live in a home built before 1978 and it has not been completely remodeled (or the paint has not been stripped), you can consider lead abatement. This needs to be done by a professional. There is no point in trying to remove the leaded paint from your home by yourself, because you will definitely expose your family to more lead in the process.

You should also think about lead exposure if your pipes are leaded or use lead solder. Water can be tested for lead content. Warm water leaches lead much better than cold, so if you think you could have lead in your pipes, use only cold tap water for drinking and cooking. Many people also suggest running the tap for fifteen to thirty seconds prior to use to wash out lead residue.

If possible, do not buy canned foods imported from other countries. The United States has banned the use of lead solder in canned foods, but imported cans may still use lead.

And ask your pediatrician to check your child's lead level if it hasn't already been done. This is a fairly easy test and can be very reassuring. Parents can be tested as well, though remember that kids absorb lead much more readily than we do and they are "allowed" a lower blood-lead level than we are. So it is much more important to screen your child than yourself. Still, it's an easy test that I have performed on many curious parents.

WHAT'S IN MY HOUSE?

I will confess that I don't think about lead very often. We rebuilt our house from the ground up a few years ago, so by definition it is lead free. The issue of lead exposure for my children simply doesn't cross my mind that frequently. But my daughter came home from kindergarten one day, reporting that she had learned to run the water fountain for a few seconds before drinking in order to "rinse out the lead." I am truly glad her teacher thought to pass along that fact to the class—because it didn't occur to me.

Chapter 15

Pesticides

WHAT IS THE QUESTION?

The term "pest" includes just about every unwanted critter you can think of, from salmonella on your uncooked chicken to a rat lurking in your backyard. Pests are everywhere, invading homes, gardens, schools, hospitals, agricultural fields, and crates of harvested produce.

Pesticides get rid of the little buggers, improving not just agricultural yields but also overall public health. But despite their benefits, pesticides have a bad reputation. The very word "pesticide" conjures an image of a toxic spray wafting through the air and carrying with it the potential for deadly disease. Though the reality is not quite that dramatic, there are known health risks to pesticide exposure, which is why we have this stereotype.

The topic of pesticides is enormous and these chemicals are all around us. The question of whether pesticides are dangerous or safe isn't illegitimate, it's just too big. So in this chapter I take a look at the smaller subset of agricultural pesticides as an example of how people can think about pesticides in general.

The story of what has happened to agricultural pesticides over the past decade is amazing. It provides a way to understand what we should worry about, what we can do to minimize our risks, and what government agencies can do (and already have done) to protect us more. Do you need to modify your exposure to agricultural chemicals? Does this mean you need to buy only **organic** foods? It might surprise you that when you go to the supermarket, you may be able to avoid pesticides without going exclusively organic.

WHAT IS THE DATA?

A pest is, by definition, any organism living anywhere it is not wanted. Some pests cause damage to crops; others inflict their harm on humans or animals. The cockroach in your cupboard is a pest. The weed in your yard is a pest. The termites living in the frame of your home are pests. The fungus destroying a farmer's crop is a pest. The mosquito spreading West Nile virus is a pest. My daughter might occasionally describe her little brother as a pest—and although he sometimes fits the technical definition (an organism living where he is not wanted), my son does not really qualify.

Pesticides are everything that people use to kill—or at least get rid of—pests. Farmers use these chemicals to remove insects and fight infections that lower their harvest yield. We use similar substances in our homes to evict critters we don't want to share the house with, everything from mice to ants to mold; we even use pesticides on pets in the form of flea treatments. We put them on our countertops (disinfectants), in our swimming pools (chlorine), and on our lawns. Even hospital sterilizing solutions and antibacterial soaps are technically pesticides. According to the EPA, if you add up all the available pesticides, you will find more than one thousand active ingredients in more than seventeen thousand registered products.[1]

The pesticide industry is still growing: there have been one hundred new chemical pesticides registered since 1997 alone. It is worth noting here that the EPA is *the* pesticide clearinghouse. All pesticides must be registered with the EPA so it can review their ingredients and potential effects on humans. When it comes to foods being

---

DECODING THE WARNINGS ON PESTICIDE LABELS

It turns out that the words on the label of your pesticide are meant to do more than caution you about the risk of the product. They are chosen specifically to indicate the level of danger if you ingest or come into direct contact with the product. This is still pretty vague, but who knew that there was a subtle meaning to each of these terms?

CAUTION means low toxicity

WARNING means moderate toxicity

DANGER means high toxicity

treated with pesticides, the EPA sets limits on how much pesticide can be used and how much residue can remain on the produce. While we think of pesticides mostly in terms of our homes, lawns, and foods, a huge number are used in hospitals and health-care facilities as well. Pesticides can be divided into classes. There are chemical pesticides and biopesticides. There are algicides, fungicides, herbicides, insecticides, miticides, molluscicides, ovicides, and rodenticides. Each of these classes has dozens—even hundreds—of agents, all of which work in different ways. There are some, like the organochlorines (made famous by DDT) and organophosphates (like parathion, malathion, and sarin), that affect the nervous system of their targets; and there are others, like the pyrethroids and the neonicotinoids, that mimic naturally occurring pesticides. I could go on with the multisyllabic scientific names and you could go to sleep.

It is tempting to categorically demonize pesticides, but it wouldn't be entirely fair. When you go to the grocery store there is usually a bounty of produce. This is largely a result of pesticide use: these chemicals minimize crop damage and improve crop yields. In fact, many believe that pesticides deserve much of the credit for the average American diet having more fresh fruits and vegetables these days, and thus improving the overall public health. Ironic, isn't it? As a nation we eat more healthy food, but that food may be covered with unhealthy chemicals. Still, pesticides are not all bad.

Pesticides used in medical settings have obvious benefits too. Disinfectants clean countertops, removing highly infectious liquids like saliva and blood. Antibacterial cleansers help to get rid of infections that may otherwise infect a cut or scrape. We don't really think of these as pesticides, but each of these products qualifies.

Likewise, pesticides in our homes make them more livable and often more beautiful. Granted, if you choose not to treat your front lawn with a greening agent that contains pesticide, it is possible that no one would notice. But the same is not true if you chose not to get rid of the ants swarming in your kitchen or the cockroaches under the bed or the rats that scatter their droppings on your living room floor. I wouldn't want to go over to that house for dinner, would you?

But pesticides clearly have their downsides. For one thing, pesticides are effective because they harm or kill bacteria, viruses, fungi, and small animals. A chemical that kills teeny-weeny defenseless organisms—how is that not scary? It is not just *that* pesticides kill; it is *how* they kill. They affect organs in insect bodies, which, though much smaller, are very similar to our own organs. Could the toxicities that make pesticides effective turn around and make humans sick too? There are entire books and Web sites dedicated to pesticide education. These are phenomenal resources able to explain everything from how particular pesticides function to the potential human side effects. This chapter is not about singling out specific pesticides or reviewing individual health claims. Rather, it asks whether pesticides—specifically agricultural pesticides used on our food supply—deserve their bad rap and, if so, what we can do (and what we have done) to protect ourselves. Ultimately, when we consider our grocery list, do we need to bite the bullet and buy only organic?

Organophosphate pesticides are the most commonly used pesticides in agriculture.[2] These chemicals work by irreversibly inactivating an enzyme (called acetylcholinesterase) that is critical for normal nerve-impulse transmission. Without acetylcholinesterase, one nerve cannot communicate normally with the next, so the signal cannot be conducted as it should be. The kicker is: acetylcholinesterase works the same way in insects and humans. So presumably, organophosphates are as toxic to our neurons as they are to pests'. Organophosphates have such a bad reputation at this point that they might as well have a skull and crossbones in place of the leading "o."

Some experts defend organophosphates on the grounds that the chemicals degrade quickly when exposed to sunlight, air, and soil. In other words, there shouldn't be much of the pesticide on your food. But if you do come into contact with a significant dose, it can be toxic. There are many cases of agricultural workers poisoned by various organophosphates.

Today's new twist on organophosphates has to do with infants and children who have developing brains: what dose of organophosphate is truly safe for them? In fact, what dose of any pesticide is truly safe for a growing bundle of neurons in your child's head?

When a pesticide is found on food, it is called residue. I always thought of residue as something that could be rinsed off, like dirt. It makes sense: if the pesticide is sprayed on the plant or even if it is tilled into the soil, it shouldn't become a part of the fruit, should it? But in every study I am about to cite, the researchers washed the produce as you

THE EPA'S LIST OF TIPS YOU CAN USE TO PROTECT
YOUR KIDS FROM PESTICIDE POISONING

1. Always store pesticides and other household chemicals, including chlorine bleach, out of children's reach—preferably in a locked cabinet.
2. Read the label *first!* Pesticide products, household cleaning products, and pet products can be dangerous or ineffective if too much or too little is used.
3. Before applying pesticides or other household chemicals, remove children and their toys, as well as pets, from the area. Keep children and pets away until the pesticide has dried or as long as is recommended on the label.
4. If your use of a pesticide or other household chemical is interrupted (perhaps by a phone call), properly reclose the container and remove it from children's reach. Always use household products in child-resistant packaging.
5. Never transfer pesticides to other containers that children may associate with food or drink (like soda bottles), and never place rodent or insect baits where small children can get to them.
6. When applying insect repellents to children, read all directions first; do not apply over cuts, wounds, or irritated skin; do not apply to eyes, mouth, hands, or directly on the face; and use just enough to cover exposed skin or clothing, but do not use under clothing.
7. Wash children's hands, toys, and bottles often. Regularly clean floors, windowsills, and other surfaces to reduce possible exposure to lead and pesticide residues.

Questions about pesticides? Call the National Pesticide Information Center (NPIC) at 1-800-858-7378 or visit online at http://npic.orst.edu/.

or I would; if there was a peel, the peel was taken off. In other words, based upon the data you are about to read there still seems to be a good load of residue hanging on to produce after rinsing. Somehow, it sticks.

Agricultural pesticides gained the spotlight in the early 1990s. If you work in agriculture or food services, you may think I am way off on that one. But for the majority of people, pesticide residue was a concept that flew way below the radar until that time. Likewise, in the early nineties the organic section in most supermarkets was the shelf with the funky-looking apples and the lettuce you never wanted to eat.

Starting in 1991, the U.S. Department of Agriculture (USDA) stepped up its effort to sample produce, looking for pesticide residue. The first set of results was widely published in 1996. In the report, the USDA identified pesticide residues on more than 90 percent of a variety of fruits and vegetables, including apples, peaches, strawberries, and sweet bell peppers. Other produce with fairly hefty pesticide residues included carrots, spinach, lettuce, and pears. The 1996 report was perhaps most significant for its documentation that these results were consistent year after year.

Now here's the catch: most of the time the residue levels were below the limit considered acceptable for exposure. In other words, when a scientist looked at how much of any one pesticide was on any one piece of produce, most of the time it was an amount that had been deemed safe by the government. But the researchers often found multiple pesticides on individual fruits. For instance, more often than not apples had more than four different pesticides, and there is one report stating that an apple sample had as many as thirty-nine different pesticide residues, fifteen of which were organophosphates. The governmental agencies reassured consumers that residue levels were safe because if each residue was considered individually it might be. But consumer and environmental groups started to protest and demanded that the residues be looked at cumulatively. It didn't matter what one of the thirty-nine residues on an apple did to my body; it mattered what all thirty-nine did while entering my body simultaneously.

In 1996 Congress passed the Food Quality Protection Act (**FQPA**). This changed the landscape by putting into action regulations based on child safety. Years earlier the National Academy of Sciences had issued a stern warning recognizing that children are more vulnerable to pesticides, especially organophosphates, because the chemicals could affect the developing brain. What's more, one- and two-year-olds are exposed to more pesticides per kilogram of body weight because their diets tend to be fruit (and sometimes vegetable) intensive and because they consume more calories per kilogram than their adult counterparts. The National Academy of Sciences begged for new standards for this age group, and the FQPA gave them what they asked for.

The new law required "an additional tenfold margin of safety" for infants and children—the minimally acceptable standards had to be ten times safer than they had been up to that point. It also said that there had to be better health-based yardsticks by which to measure the safety (or danger) of pesticide residues. The people setting these standards had to prove that at a given residue level there was "a reasonable certainty of no harm" to the consumer. And for the first time, when considering consumer safety the EPA was required to factor in other sources of pesticide exposure, like drinking water and residential pesticides.

The USDA came up with a score called the Dietary Risk Index (**DRI**), which was calculated from residue levels found on a particular food and a pesticide's specific toxicity. The first DRI scores were reported in 1994. Apples had a DRI of 455, tomatoes 452, sweet bell peppers 587, and so on. The USDA has since continued to repeat the study and calculate DRI scores year in and year out.

The new standards were voted into law in 1996. A study by the Environmental Working Group around the time the law was passed illustrates the starting point. Researchers found that a massive number of children in the United States between one and five years of age ingest pesticides every day: twenty million of them consume eight different pesticides each day. Looking by particular types of pesticide, more than six hundred thousand kids were eating doses of organophosphates technically considered neurotoxic.[3] This was the baseline. The FQPA required that within ten years the standards would be changed and the problem would be well on its way to being fixed.

By 2003 the impact of the FQPA reported by the government was dramatic. Some of the produce—like green beans, broccoli, and grapes—saw a nearly 100 percent decrease in pesticide residue. Apples were down 97 percent, and tomatoes 85 percent. Some produce benefited less dramatically, but the vast majority saw DRI scores fall well below 100. By 2006 it was estimated that the risks associated with sixteen foods commonly eaten by children had decreased by 50 percent.

If you believe these numbers—and, frankly, I do—the federal Food Quality Protection Act is an amazing example of a law working swiftly to improve our health. By shifting the standards to better protect children, we all benefit. And in ten years, there has been a huge reduction in pesticides found on store-bought fruits and vegetables. Thin-skinned or fleshy produce still tends to have higher pesticide levels than thicker-skinned varieties, but this makes logical sense. Pesticides have a hard time penetrating a thick outer layer, so bananas are better protected than strawberries.

Here's another example of how the FQPA seems to have made a difference. The law only applies to domestic produce. As a result, today imported fruits and vegetables tend to have higher pesticide levels than those grown conventionally—not even organically—in the United States. Seems counterintuitive, doesn't it? We are so predisposed to think of this country as the biggest offender on all environmental fronts, but not so when it comes to pesticides on produce.

If you read the chatter on agricultural Web sites and blogs, the FQPA is mentioned constantly. Some groups are certain that pesticides are well on their way to being a thing of the past. Farmers describe pressures to grow their produce without pesticides because it's just a matter of time before everyone will be required by law to do so.

Of course, there is always another side. First of all, not everyone believes the data released by the USDA. There are dozens of "independent" studies—some legitimate, others not so—analyzing various fruit and vegetable pesticide residues. The Northeast Organic Farming Association of New York, for instance, lists on its Web site twenty conventionally grown products still replete with residues. The site also quotes a study from the University of Washington School of Public Health that shows children who eat conventional produce have nine times the organophosphate levels in their urine than children who eat organic produce. But here's the problem: we are not comparing apples and apples (pardon the pun). It is true that there are still pesticide residues on conventionally grown produce— neither the EPA nor the USDA would argue otherwise. The question at hand is what do these residues do to our children's bodies (and our bodies)? By extension, have the residue levels gone down to a more acceptably low level?

It seems to me that unless pesticides are outlawed, there will forever be an impasse here. The government agencies are lowering the acceptable risk level, but until that level goes to zero there will be a group of people who don't think the government has done enough.

And here's where I sound like I am talking out of both sides of my mouth: the concerns expressed by the pesticide critics are reasonable.

EWG'S DIRTY DOZEN

The Environmental Working Group (EWG) tabulates a list of produce with high pesticide residue levels. The big offenders include peaches, apples, bell peppers, celery, nectarines, strawberries, cherries, kale, lettuce, grapes, carrots, and pears. (From http://www.foodnews.org)

These people argue that at different doses pesticides have been linked with all sorts of illnesses, like diseases of the nervous system, lungs, reproductive system, and immune system, as well as cancers.[4] Critics look at the data on organophosphates and voice strong concerns about how neurotoxins affect developing brains. They invoke newer data showing that some pesticides may act as **endocrine disruptors**, affecting the way hormones work in some people's bodies. They bring up the fact that these days girls tend to enter puberty at an earlier age, and some postulate that this is a result of endocrine-disrupting effects of pesticides. They discuss the rapid increase in food allergy, and some suggest that the reactions are to pesticides rather than to the foods carrying the pesticides. They bring up genetically modified plants, engineered to withstand pesticide spraying, and ask whether we have considered the health implications of introducing new genes to our food supply. All of these, actually, are good questions. All of them deserve study so that we can get to the answer.

In the meantime, while we are waiting for the data, these pesticide opponents argue that any exposure is too high an exposure. I think it is important to recognize that this approach—if we don't know what it does to our body, we ought to avoid it—is rational. But rational isn't always practical. When you go to the market, organic produce can be two or three times the cost of conventionally grown fruits and vegetables. Some stores carry a wide selection of organic products, but most offer only a few. If you face the choice between giving your children fresh fruit that is conventionally grown and may have a small amount of pesticide on it versus not giving them any at all, I would argue that the health benefits of the fresh fruit are more important. The alternative is often processed food made with refined sugars and a high fat load, and we know the implication of that kind of diet: it causes obesity, diabetes, heart disease, and a slew of other long-term health repercussions. Compared to the theoretical risk of pesticides, I choose the fresh produce. Frankly, unless the government decides to put the pesticide industry out of business, it doesn't look like zero pesticide exposure is on the horizon.

To add one more complicating twist, the word "organic" doesn't necessarily mean one specific thing. The National Organic Program is the arm of the USDA responsible for setting and implementing organic standards. Within the general category of organic foods, there are levels of certification:

- "100 percent organic" means that all ingredients and processing aids are completely organic.
- "Organic" means that at least 95 percent of the ingredients are organic.
- "Made with organic ingredients" means that at least 70 percent of the ingredients are organic.

The last two categories are "products with less than 70 percent organic ingredients" and "organic livestock feed." So even if people do bite the bullet and decide to buy organic groceries, they need to read the label. An "organic" food may have 5 percent of its ingredients coming from nonorganic sources, and that can equal pesticide exposure. A food "made with organic ingredients" can have up to 30 percent of its ingredients nonorganic.

As I said before, I am inclined to believe the government's data. The FQPA has made a tremendous dent in the issue and is probably responsible for some of the mainstreaming of organic foods today. When the government gave itself the charge to reduce pesticide residue exposure, conventional farmers complied. Organic farmers, meanwhile, looked like geniuses, and the crummy organic produce row grew into massive, elegant chains like Whole Foods.

We don't really know what pesticides in low doses do to any of us, ourselves or our children. The health effects of pesticides in food clearly need to be better studied. But if we take a giant step back and look at the broader issue of pesticides, we are really asking about the general public health risk associated with pesticide use. Often it is a risk that we don't choose, because it is an unknown chemical exposure. Once a farmer tills pesticide into his soil, it's there. If a school janitor sprays the classroom with disinfectant, the doctor washes her hands with antibacterial soap, or the fumigator fogs an office building, the chemical is out there. You haven't been asked if it is okay, and often you don't even know that it has been done.

No one can control all of life's potentially dangerous exposures, pesticides included. But what we can control is what we buy at the market. If you can get organic produce—especially the thin-skinned or naked varieties that will obviously collect pesticide residue on their edible fruits or leaves—then buy it. But if you cannot afford only organic, or if you cannot find only organic, I wouldn't lose sleep. Do the best you can. Even when you are buying conventionally grown domestic produce, you are eating many fewer pesticides today than you were a decade ago.

WHAT IS THE BOTTOM LINE?

Organic produce has fewer pesticide residues than conventionally grown (nonorganic) foods. Nobody debates that. So if you can buy organic fruits and vegetables—if they are affordable and available—then go for it. Conventional produce will continue to have at least some residues until pesticides are no longer used in conventional growing. Organic foods may have a few pesticides of their own, but the amount of residue is clearly less.

The real question is this: are there few enough pesticide residues on nonorganic produce (grown in the United States) that it is probably safe? In many cases, the answer is yes.

Meanwhile, there is still much more work to be done. Children are uniquely vulnerable to potential harm from pesticides on produce because pound for pound they eat more than their adult counterparts. This means that an average child is exposed to a higher load of pesticide residue than an average adult. Little kids—infants and toddlers—play on the ground and in soil, putting their hands in their mouths and picking their noses. This exposure is not insignificant: small children eat their fair share of pesticides (and dust and bacteria and so on) without even ingesting any food.

The question of whether pesticides are dangerous or safe is impossibly broad—it is really seventeen thousand questions, given the number of pesticide products available. What one particular pesticide does to a person depends upon the chemicals involved, the dose, the amount absorbed by the body, and so on. But that said, the federal Food Quality Protection Act has gone a long way to minimize consumer exposure to residues in agricultural produce. Steps have been taken without knowing the exact biological effects of pesticide residue. That's pretty progressive.

WHAT'S IN MY HOUSE?

I always assume that pesticide exposure is one of the risks that go along with eating an apple. If I choose a smaller, mealier looking piece of fruit, I may be selecting one with less pesticide exposure and less risk. I might prefer that perfect green, slightly oversized, super-shiny apple. I just deserve to know the potential harm, weigh the pros and the cons, and choose whether or not to minimize my risk.

In general, I have moved toward buying more organic produce. But I really focus on the thin-skinned and "naked" foods. I just can't wrap my brain (or my wallet) around buying organic bananas quite yet. With the results of the FQPA, I don't think I'll ever really need to.

Chapter 16

Plastics

WHAT IS THE QUESTION?

Interest in plastic bottles started with environmental activists. They were the first to bring the debate front and center, weighing the convenience of single-use plastic bottles against the costs to the environment. Environmentalists have long said that the ease of grabbing a bottle on the run is outweighed by the burden involved in the disposal and recycling of those bottles. Millions of disposable plastic containers are adding up to a small global nightmare. But recently, the focus on plastics has shifted. Instead of talking about their ecological burden, people are worried about the threat plastics pose to individual health.

We schlep around a water bottle everywhere we go in an effort to drink more throughout the day, because supposedly the more water we drink, the healthier we are. The bottle is invariably left in the car or on your desk for a few hours at a time. As it warms up in the light of the sun, small beads of water-sweat line the inside neck of the bottle. You open the top and take a gulp because you know you should be hydrating to keep healthy, but you can't help but taste the chemical flavor of the now-warmed water. Does the very plastic that contains your water contaminate it?

The chatter over plastics broadened from single-use to multiple-use plastic containers and now also includes just about anything made with plastic containing **bisphenol A**—from baby bottles to the linings of food and beverage cans. Researchers and consumers alike have become increasingly concerned about the effects of this chemical on our bodies.

Is that plastic container actually contaminating your drink? Should your baby use a plastic bottle? Can you eat canned foods if the lining contains bisphenol A? How toxic is plastic, really?

WHAT IS THE DATA?

On the bottom of all containers made of plastic is imprinted the universal recycling symbol—a triangle made from three arrows—with a number stamped inside (also called the "resin identification code"). The symbol tends to be so small that it is nearly impossible to read the resin code. The system of coding was created to help recycle plastics: each number represents a different type of resin used to make a given plastic, and the method of disposing and recycling each of these is different. Despite rumors to the contrary, these codes do *not* rank the relative safety (or danger) of various plastics.

The key to the code is as follows (briefly summarized from the American Chemistry Council's "Plastic Packaging Resins" chart found on its Web-based learning center):

♻**1** **PET** (polyethylene terephthalate): clear, smooth, and tough plastic; used for shatterproof beverage bottles and other containers like peanut butter and pickle jars; can be recycled into fiberfill to make sleeping bags, carpet, rope, and tote bags.

♻**2** **HDPE** (high-density polyethylene): stiff bottles made to be translucent or pigmented; good for products with a short shelf life (like milk); also good for household and industrial chemicals like detergents or bleach; used as cereal box liners and butter tubs; can be recycled into trash cans, plastic lumber, pipes, flowerpots, traffic barrier cones, and detergent or cosmetic bottles.

♻**3** **PVC** (polyvinyl chloride): can be fashioned into rigid or flexible plastics; rigid uses include blister packs (medicines), pipes, window frames, decking, and so on; flexible uses include shrink-wrap and tamper-resistant wrapping as well as medical (blood bags, tubing) and household (carpet backing, flooring) uses; can be recycled into many of the household supplies for which it is designed, including pipes and flooring.

⟳⁴⟲ **LDPE** (low-density polyethylene): tough, very flexible, and transparent; used as shrink-wrap and in a range of bags (dry cleaning, newspaper, garbage); as coating for beverage cups and lids; to make squeeze bottles and toys; can be recycled into garbage can liners and trash cans, tiles, and outdoor lumber.

⟳⁵⟲ **PP** (polypropylene): strong with a high melting point, so it is good for hot liquids; used in containers (for yogurt, take-out foods, and deli foods), medicine bottles, and large molded automotive parts; can be recycled into plastic lumber, automotive materials like battery cases and oil funnels, and manhole steps.

⟳⁶⟲ **PS** (polystyrene): usually used in its clear, hard, and brittle form but can also be foamed; used in disposable food service items (cups, plates, take-out containers, etc.), packaging materials (also called "peanuts" or "loose fill"), CD cases, and other electronics housings; can be recycled into plastic moldings, insulation, and cartons or protective packaging.

⟳⁷⟲ **Other:** this group contains a resin other than one of the six listed above or a combination of two or more of the resins above; properties depend on what resin is used, but the most common uses are reusable water bottles, citrus juice and ketchup bottles, oven baking bags, and custom packaging; can be recycled into bottles and plastic lumber.

Currently, the most publicly controversial component of all plastics is bisphenol A, also known as **BPA**. Technically, plastics containing BPA fall into the "other" (no. 7) category.[1]

BPA is a chemical found in hard, clear polycarbonate plastics. It has been used by the plastics industry since the 1950s to make products like sports equipment, medical devices, CDs, and home electronics. It is also used to coat the inside of some canned goods; as a precursor to a widely used chemical flame retardant; in dental sealants; and, of course, in the manufacturing of multiuse water and baby bottles. It is worth noting that polycarbonate is *not* used in the manufacture of one-time-use containers like soda bottles or disposable water bottles. Rather, it is used in reusable containers like baby bottles and the original Nalgene bottles.

As these plasticized products have become a ubiquitous part of our society, BPA exposure has multiplied. In fact, recent studies in the United States show BPA in the blood or urine of 90–95 percent of all people tested (Calafat 2008). We don't just eat and drink BPA; we also breathe it in the form of house dust and absorb it through our skin when it comes into direct contact with us.

BPA falls into the category of **endocrine disruptors**, chemicals that appear to interfere with hormones and other signaling systems in the body. Specifically, BPA is accused of mimicking the effects of estrogen. Studies have shown that exposure to high levels of BPA causes increased risk of uterine fibroids, endometriosis, and breast cancer; it lowers sperm count; and it increases the risk of prostate cancer. Each of these diseases is sensitive to hormone effects, many implying estrogen-like actions of BPA. But these are one-time studies or involve small sample sizes, so critics question their results.

Some researchers believe that the effects of BPA in the body go beyond imitating estrogen. Endocrine disruptors can potentially affect the way different hormones work all over the body. Along these lines, there is emerging data that BPA may cause insulin resistance and ultimately contribute to obesity (Alonso-Magdalena 2006). Insulin is the chemical in the body responsible for helping glucose (sugar) get into cells so it is can be used as energy. When the cells become resistant to insulin, they cannot take in glucose. The glucose sits around in the bloodstream, building up to excessive levels. High blood sugar is commonly known as diabetes.

In September 2008, new data suggested that BPA might even affect the body in ways that have little—maybe nothing—to do with hormones and endocrine disruption. Here high BPA levels were implicated in cardiovascular disease (including heart attacks) and liver disease (Lang 2008).

Understanding BPA's effects in infants and children is unbelievably difficult. Kids are generally harder to study than adults to begin with— parents tend to be reluctant to enroll their children in studies. This is further complicated by the fact that these days exposures begin in the womb. In 2007 the National Toxicology Program (NTP), a branch of the **NIH**, published a statement by several experts raising concerns about even low levels of BPA in infants and children. After analyzing several hundred studies on BPA, they concluded that the chemical might affect the nervous systems and the behaviors of fetuses, infants, and children. They also wondered out loud if BPA is responsible for the earlier onset of puberty in females (Chapin 2008).

So it seems that BPA is everywhere and its potential impact on our bodies keeps broadening, now including organs like the heart, liver, pancreas, brain and nervous system, uterus, and prostate. How much BPA exposure is safe?

The EPA has currently recommended that people get no more than 50 *ug* per kilogram of body weight per day of BPA exposure. For an average nine-year-old child weighing sixty-six pounds (30 kg), this equals 1,500 *ug*/day; for a newborn weighing seven pounds (3.2 kg), it is only 160 *ug*/day. The problem here is that it is impossible to know how much is coming at you or your child on any given day.

Furthermore, many now question this recommended limit. The 2007 NTP report claimed that BPA harms animals at low levels—levels found in nearly all human bodies. It also suggested that infants and young children face much higher exposure levels than adults just given their lifestyle: BPA seeps from plastic baby bottles and leeches from liners in baby food and infant formula cans.

BPA is leached most effectively by pouring boiling hot liquids into polycarbonate drinking bottles—BPA levels in the boiling liquid rise up to fifty-five times faster than in cool or temperate water (Le 2008). But take a breath here. Rarely, if ever, do most parents actually boil liquids that they are preparing for their children, and even if some moms and dads do, they almost never put that boiling water into plastic bottles before letting it cool. If you do, this is one habit worth changing.

The dishwasher also causes an increase in BPA leaching because the hot water cycle followed by heated drying can age plastic quickly, resulting in cracks along the lining. These cracks provide more access to BPA when the bottle is filled with milk or water. For this reason, I have always been a fan of hand washing and rack drying bottles.

In April 2008, the Canadian government decided that BPA posed enough of a risk to infants that it classified the chemical as "toxic" to human health and to the environment. This further catapulted the arguments of the anti-BPA camp in the United States.

Since the first printing of this book, the EPA and the Department

MICROWAVING

Is it dangerous to microwave food in plastic containers? Parents are warned against microwaving formula or breast milk in plastic bottles. However, this warning did not initially come from fears of leaching BPA. Rather, it was put in place to prevent burns. When food or liquid is microwaved—and then not thoroughly shaken or stirred—there can be "hot pockets" with boiling temperatures. The real intent of the warning was to avoid burn injury to babies from hot pockets. Now, with data about increased BPA leaching by boiling liquids, there is another reason not to do it.

Microwaves have been on the receiving end of a lot of bad press in the debate over BPA, with many people assuming that a food or liquid heated by a microwave while in a plastic container must have elevated levels of all sorts of chemicals leached from that container. However, as this explanation from the *Harvard Medical School Family Health Guide* (2006) explains, this assumption is not accurate:

> The FDA tests measure the migration of chemicals at temperatures that the container or wrap is likely to encounter during ordinary use. For microwave approval, the agency estimates the ratio of plastic surface area to food, how long the container is likely to be in the microwave, how often a person is likely to eat from the container, and how hot the food can be expected to get during microwaving. The scientists then measure the chemicals that leach out and the extent to which they migrate to different kinds of foods. The maximum allowable amount is 100–1,000 times less per pound of body weight than the amount shown to harm laboratory animals over a lifetime of use. Only containers that pass this test can display a microwave-safe icon, the words "microwave safe," or words to the effect that they're approved for use in microwave ovens.

So when a plastic container says "microwave safe," this is what it means. (Keep in mind that BPA is not mentioned in this definition.)

of Health and Human Services (HHS) have shifted their stance on BPA a bit. Both acknowledge that while the FDA determined in 2008 food-related materials containing BPA were safe, they also acknowledge more recent studies in lab animals suggesting subtle effects of low dose BPA. Though there is no proof that BPA harms children or adults, the new studies have created some concern about the safety of the chemical. As a result, HHS has announced $30 million of funding for studies by NIH, CDC, and FDA over the next one to two years.

Now, there is another side to this issue. There are claims against BPA that have already been disproved, at least according to the NTP. Birth defects and miscarriages are *not* caused by BPA exposure during pregnancy. Neither is low birth weight or delayed growth in babies exposed to BPA in the womb.

The other feather in BPA's cap is that high doses may be harder to come by than once thought. The NTP worries about low-dose exposure for the young but not for adults. Researchers argue that aside from adults with high occupational exposures, most adults are not at risk for BPA-related adverse effects because the exposure levels are too low.

Basically, each and every study of BPA ultimately concludes that more research needs to be done. The 2008 study that tied BPA to heart disease, liver enzyme abnormalities, and diabetes did not sway the **FDA**, with the organization expressly stating that it still considered BPA safe.

The data on BPA is constantly evolving. It seems that the more systems researchers investigate (heart, liver, brain), the more damning evidence they find. But as of this moment, no single study is compelling enough for the FDA to pull the chemical out of our foods, drinks, and household luxuries. Even if the FDA did find convincing evidence, removing all sources of BPA in order to have zero exposure is essentially impossible. According to most experts and researchers in the field, the possibility that BPA can do harm to humans—causing disease in adults or impacting fetal development—has not been ruled out. But it hasn't been proven either.

WHAT IS THE BOTTOM LINE?

If BPA is a chemical that can potentially cause disease, why not try to avoid it? When you can, opt for an alternative to plastic. If you are at home, fill a glass with tap water rather than grabbing a plastic bottle of water. It is better for the environment and likely better for you.

If you have an infant, BPA-free bottles are the way to go. You could use a glass bottle (BPA free by definition) or try a plastic version that advertises its lack of BPA. These are made of polyamide plastics instead of polycarbonate plastics. I cannot promise that there won't be a new worrisome chemical that is discovered in these BPA-free plastic bottles in the future—since those bottles are still made of plastic, chances are there will be skeptics lining up. But at this moment, it seems logical to at least try to avoid the chemical at the center of all this controversy.

WHAT'S IN MY HOUSE?

I don't go crazy on this subject—my house is by no means a BPA-free environment—but where there is a BPA-free alternative, it seems rational to choose it. My kids drank out of BPA-lined bottles, but it was during a time when BPA was a relative unknown and I haven't lost any sleep over it. I am a believer in all things in moderation. I have not thrown out every single plastic cup and container in my home. Instead, I simply keep hot food and drinks away from plastic. I also hand wash and rack dry plastics rather than machine washing them. I have begun to take the lids off my to-go coffee cups (why sip through a plastic lid unnecessarily?), and I never cover microwave food with plastic wrap. But I don't try to read the resin code on the bottom of recyclables; frankly because my eyes aren't good enough to see them. Ultimately, the issue of BPA all comes down to common sense. If at a particular moment I can avoid plastics I will, but when I am on the go and in a pinch I use plastic containers, as do my kids.

Part IV

On Your Body

Chapter 17

Cosmetics

Nail Polish, Hair Products, and Perfume

WHAT IS THE QUESTION?

Cosmetics include pretty much all the things we use for beautification: shampoo and conditioner, makeup, nail polish, and perfume. Women are the primary consumers of cosmetics, but men (aftershave, cologne, sunscreen) and children (bubble bath, baby lotion, diaper cream) use them every day too.

People don't generally think of cosmetics as harmful because they go on us, not into us. But various cosmetics have found themselves on the receiving end of criticism—or at least controversy—over the past few years because, as it turns out, their chemical additives, like **phthalates** and **parabens**, are absorbed by our skin.

Phthalates add flexibility to products. They help nail polish avoid cracking; they help hair spray hold your locks in place without creating too much stiffness; and they maintain smell in perfumes and shampoos. They do all of these things because they are "plasticizers." This plasticizing ability also makes phthalates useful in thousands of other everday products, including vinyl flooring, toys, detergents, food packaging, and more.

Parabens are preservatives. By making it nearly impossible for microorganisms to live in and on our cosmetics, parabens extend the shelf lives of many of these products.

Here's the concern: are the phthalates and parabens in our cosmetics making their way into our bodies? And if so, do these chemicals make us sick?

WHAT IS THE DATA?

Let's start with phthalates. These chemicals were invented in the 1930s. Phthalates were not the first plasticizers—seventy years earlier inventors added camphor to nitrocellulose to make celluloid, the first "thermoplastic." Early on, celluloid was used to replace ivory (for instance, in the manufacture of billiard balls) and in the production of film. Phthalates, in fact, were designed and marketed to replace this first generation of plasticizers, specifically as a nontoxic alternative. (Oh, how things change!)

Today phthalates are everywhere. There are many different types, found in thousands of everyday commodities: food and food packaging, water bottles, cosmetics, toys, building materials, lubricating oils, solvents, medical equipment, pill coatings, vinyl flooring, carpets, paints, glue, insect repellents, and more. The phthalates add flexibility and resilience to each of these products. This chapter focuses on the phthalates in cosmetics. Though the phthalates used in other capacities are similar, the goal here is to look at whether putting phthalates on your skin or scalp causes harm. The most common ones used in cosmetics include dibutyl phthalate (DBP), diethyl phthalate (DEP), and dimethyl phthalate (DMP).[1]

In the microcosm of cosmetics, different phthalates are used for different purposes. Generally, DBP is found in nail polishes and perfumes while DEP is used in deodorants, fragrances, moisturizers, and lotions. DMP is in hair sprays.

Recently, there has been lots of publicity about phthalates in children's products. In early 2008, a study done out of the University of Washington concluded that children had higher phthalate levels if parents used "baby" shampoos, lotions, and powders containing them (Sathyanarayana 2008). This confirmed prior studies showing that the body absorbs phthalates when they are used on the hair or skin (Koo 2004). But what made headlines in this particular study was the conclusion that the absorption of phthalates varies by age. Children exposed to phthalate-containing products had twice the urine phthalate levels of kids who weren't exposed to these products; when researchers looked at infants under eight months of age, the urine phthalate levels jumped to five times those of unexposed babies. The author's conclusion was that phthalates are in fact absorbed at higher rates than once thought, and because kids don't absolutely need these phthalate-containing products, it is reasonable to avoid them.

There is a saying in research—and this includes all research, not just medical research—that "correlation is not causation." This statement is extremely important. It cautions researchers to make sure they can prove cause and effect. If you are looking at a specific chemical and trying to explain a certain disease, you have to be able to demonstrate that the chemical *causes* the disease. A high level of the chemical doesn't necessarily mean that it is the culprit. The Sathyanarayana study did not establish an association between the use of phthalates in children's cosmetic products and a specific health risk. All the study did was demonstrate that if you used a phthalate-containing shampoo or lotion on your child, the by-products of phthalates showed up in the child's urine.

So this study tells us that the body does in fact absorb these chemicals, and it absorbs them (or at least urinates them out) at a higher rate in infants. Separate studies have looked at how phthalates make their way into our bodies—they can be absorbed through the skin, inhaled through the mouth or nose, ingested orally, and even enter the bloodstream directly through intravenous transfusions and medical devices like IV tubing. But none of these studies say anything about what the phthalates do once they are inside. What is their biological effect? Our exposure to phthalates is chronic; over their lifetime, our children will be exposed in much higher quantities than we ever were. If phthalates do cause harm, it is important to figure out how and why.

In trying to solve the phthalate mystery, researchers have honed in on hormonal effects. Phthalates seem to affect reproductive system development and hormone levels in males. It is interesting to see males on the receiving end of a chemical assault because most of the feared chemicals in our world seem to affect females and their hormones. But phthalates appear to target male hormones called androgens.[2]

Most of this data comes from studies on DBP. In 2002 the CDC published a study suggesting that DBP exposure is higher than many people expected, and it is particularly high in women of childbearing age. Researchers repeated the study in 2005, with similar results. In the second study, data demonstrated that children ages six to eleven have higher concentrations of phthalate metabolites in their urine than people over age twelve. DBP is of particular interest because animal studies show that rats fed high doses of DBP develop benign testicular tumors called adenomas. There is also some data suggesting that exposure to DBP causes shortening of the distance between the base of penis and the anus in newborn boys. DBP has therefore been added to the growing list of endocrine disruptors (**EDs**), or chemicals that affect the action of hormones in the body. Most of the endocrine disruptors (like the **phytoestrogens** in soy or **BPA** in plastics) seem to target estrogen. DBP, however, acts against the male androgens. Putting it all together, the levels of DBP are highest among children and women of childbearing age, and this chemical exposure seems to target males preferentially by affecting their genital development and predisposing some to benign testicular cancers. What this implies for future health is still up for debate.

When it comes to endocrine disruptors like phthalates, the United States seems to take an approach that is quite opposite to that of some of our international colleagues. In 2004 the European Union banned DBP and other cosmetic ingredients because of their association with endocrine disruption, cancer, and possible birth defects. The **FDA**, however, doesn't act on theories and possibilities. Instead, the agency has made it clear that in order to enact a policy change there must be sufficient and reproducible data making a case against the substance in question. One friend of mine describes it this way: in many parts of the world a chemical is guilty until proven innocent; in the United States chemicals are innocent until proven guilty.

Many have criticized the FDA for this approach, saying that it results in sluggish change and unacceptable exposures. But the FDA continues to hold strong. So when it came to phthalates, the FDA specifically stated that CDC data showing higher than expected levels of phthalates in urine did not prove an association between the chemicals and disease. As a result, the FDA simply concluded that there is more to learn about how these chemicals affect humans but there is no reason to change their current policy.

The FDA oversees the safety of food, drugs, and cosmetics. It is governed by the federal Food, Drug, and Cosmetic Act (**FD&C Act**) of 1938, which defines cosmetics as products for "cleansing, beautifying, promoting attractiveness, or altering the appearance." To be considered a cosmetic, a product needs to fit just one of those criteria. Take note that cosmetics are *not* considered drugs and therefore they are regulated differently—much less stringently—than medications. The only cosmetic ingredients requiring FDA safety testing and approval are color additives (other than coal-tar hair dyes). Therefore, testing and approval of cosmetic ingredients like phthalates is quite lenient.[3]

Outside the world of cosmetics, phthalates are subject to completely different safety and regulatory rules. For instance, phthalates are considered air and water pollutants; some are even classified as hazardous waste. It is kind of amazing that phthalates are judged so differently depending if they are in food, cosmetics, air, water, or consumer or medical products.

Many people worry that phthalates in cosmetics are not subject to rigorous enough study. A common criticism is that phthalates are studied individually—DBP is studied independent of DEP and DMP and so on. But phthalates don't work in our body in isolation. In fact, we are often exposed to multiple types of phthalates at once, and their effects may be additive or even synergistic. According to the Environmental Working Group, 84 percent of the U.S. population has six or more different phthalates in their body at any given time.

It is difficult to know when you are being exposed to phthalates in the first place, let alone how many types and which specific ones. The law requires that phthalates be disclosed on the ingredient list of retail cosmetic products. But there are exceptions for perfumes or fragrances, including fragrances used within cosmetics, like the scent added to a shampoo. Phthalates also don't have to be listed on cosmetics for professional use, nor do they have to appear on the label if they are part of a trade secret. So the law splits hairs here, demanding that the chemical is put on a label but only when it is used in a very specific way.

In 2007 California bucked that law by passing the California Safe Cosmetics Act. This bill requires cosmetic companies to report any potentially harmful ingredients to the Department of Health Services, which in turn will alert consumers. The California bill is being used as a model by many other states, and similar legislation is expected to pass elsewhere.

Meanwhile, no one knows what these chemicals do to us. There are no studies proving that cosmetics cause malignant cancer. Studies out of the **NIH** have concluded that phthalates pose minimal or no reproductive risks. Some animal studies have linked phthalates to birth defects in the male reproductive system (like undescended or absent testicles), but other studies refute this, saying that even if the phthalates were absorbed by the body they are broken down so quickly that there is no way they could exert toxic reproductive effects. The only reproduced data is that phthalate exposure in utero causes a shortened distance between the base of the penis and the anus. No one—including physicians, scientists, and researchers—has figured out the significance of this shortened distance.

One place where phthalates have clear clinical effects, however, is the lungs. Over time, everything in your house (and office, car, and so on) breaks down. With wear and tear, the materials that make up your furniture or your bedding become components of house dust. Even parts of our own bodies, like dead skin cells and shed hair, are found in the mix of a dust bunny. Phthalates are no exception. When phthalates are inhaled in the form of house dust, children experience an increase in asthma and allergic rhinitis. Mind you, other components of dust and even house dust in and of itself can trigger asthma and allergies, so the additive effect of phthalates needs to be quantified. But this is the one scenario where researchers agree on a biological phthalate effect.

As if all of this controversy over phthalates weren't enough, there is another group of chemicals in our cosmetics that garners just as much attention. These are the parabens. On the label, you may see them listed as methylparaben, propylparaben, ethylparaben, butylparaben, or benzylparaben.

Parabens are preservatives that make your cosmetics last longer. These substances create an environment that is difficult, if not impossible, for a microorganism like a bacteria or fungus to survive. The general idea is no different from using salt to cure meat or a refrigerator to chill bread. In both these cases, the preservative (salt or fridge) slows the growth of microorganisms, keeping your food fresh longer. Chemical preservatives were designed with the same idea in mind.

Unfortunately, parabens have earned a bad reputation. Yes, your bottle of shampoo or hair spray gets you through weeks—if not months—but some people contend that the very preservatives that help maintain shelf life may cut our own lives short. Do parabens cause cancer?

In 1984 the Cosmetic Ingredient Review (**CIR**) published results about three specific parabens: methylparaben, propylparaben, and butylparaben. The CIR is a cosmetics-industry-sponsored organization that reviews cosmetic ingredient safety and then publishes its results in the *International Journal of Toxicology*. The FDA also participates in the CIR, though in a nonvoting capacity. The 1984 report concluded that these parabens were safe for use in cosmetic products at levels up to 25 percent, whereas most parabens are used at a fraction of that level (usually on the order of 0.01–0.3 percent). So the FDA took this information and developed its rationale for cosmetics safety: basically, the parabens in cosmetics are safe.

Of course, because cosmetics are not considered drugs according to the FD&C Act, there are less-stringent criteria for testing and government approvals. Similar to the debate over phthalates, parabens in cosmetics are governed by different (lesser) regulatory requirements than the medicines in your medicine cabinet. Remember, other than color additives, cosmetic ingredients don't require FDA approval.

Over the last twenty years, consumers have grown increasingly vocal. They want to know what all these chemical additives like phthalates and parabens are doing to our bodies. Unfortunately, it often feels like the most vocal consumers are accusing government or pharmaceutical companies of gigantic conspiracies. As a result, these consumers are pigeonholed as those crazy few who are sure that every chemical is a toxin. In fairness, most of the consumer advocates don't fall into a radical group, and they aren't crazy. Rather, they understand that there is very little regulation in certain parts of the market (dietary supplements, vitamins, and cosmetics among them), and they simply want to know whether something they are ingesting or absorbing is dangerous or safe.

In the case of cosmetics and parabens, public pressure mounted through the end of the twentieth century. As a result, in 2003 the CIR reopened its study of parabens, and this time researchers added ethylparaben to the list of chemicals studied. After a two-year review process, the CIR decided that their original conclusions were right and there was no need to change them: parabens were safe.

At the same time that the CIR was rereviewing parabens, and a year before it redeclared them safe, a study was published in the *Journal of Applied Toxicology* showing that parabens are present in some breast tumors (Darbre 2004). This study went a step further, suggesting not only that parabens are physically present in the tumors, but also that they can act as estrogen mimics, potentially affecting breast cancer progression directly.

Not everyone agrees with this study, and follow-ups are in progress. But the public caught wind of it, and women in particular became especially suspicious of parabens, namely, antiperspirants with preservatives. An e-mail had already been going around (and around and around), warning women to stay away from antiperspirants because of a possible association with breast cancer. There was mounting concern that applying the chemicals on the skin of the armpit resulted in absorption at that spot and cancer formation in the nearby breast tissue of the upper, outer quadrant. The parabens study just fanned the flames.[4]

The American Cancer Society, the National Cancer Institute at the National Institutes of Health, the FDA, and others adamantly disagree that parabens cause cancer. These organizations cite prior and ongoing studies and continue to pursue the topic. Unfortunately, some people are still certain that there is a conspiracy of sorts: that these governmental agencies have an agenda, that pharmaceutical companies are too powerful a lobby, and that health is not at the top of the priority list. This thinking is not just wrong but harmful. Hundreds of scientists devote their lives to the study of chemicals and cancers to make your life better or longer. To dismiss their research out of hand just because it comes from a government agency is ludicrous. If there is in fact a link between parabens and cancer, the parabens lobby isn't powerful enough to keep it under wraps and these researchers are committed to letting you know. If data emerges showing a clear link, it will be published and advertised. There's no conspiracy; it's just that so far the data does not prove cause and effect.

Ultimately, there are really dozens of chemicals in cosmetics—above and beyond phthalates and parabens—that have been on the receiving end of bad press. Formaldehyde (a preservative) and toluene (a solvent) are both thought to be allergens present in nail polish. The preservatives quaternium 15 and bronopol, commonly used in baby products, break down to form formaldehyde. Coal tar, which is used in black and brown hair dyes, is a recognized carcinogen in experimental animals. Studies show that regular use of coal tar is associated with non-Hodgkin's lymphoma, Hodgkin's disease, and multiple myeloma, all relatively rare cancers but cancers nonetheless. Cosmetic-grade talc has also been shown to be carcinogenic in experimental animals. And lanolin is of concern because it can be contaminated with DDT and other pesticides. Each of these ingredients poses potential risks, and if we are going to look at cosmetics under a microscope and analyze individual ingredients, then everything on this list deserves the same attention as phthalates and parabens. There are in fact researchers studying every one of these, and in each case the answer seems to be about the same: as of now there is no clear cause-and-effect relationship between any one ingredient and serious disease.

WHAT IS THE BOTTOM LINE?

There is evidence that phthalates may act as endocrine disruptors, and endocrine disruptors may affect the way hormones work in some bodies. There is also evidence that parabens are absorbed through the skin, but scientists have yet to find a clear link causing disease. That's about all we know.

The debate over phthalate and paraben safety is not unreasonable—if a plasticizer or a preservative added to your nail polish or shampoo is not necessary and if it may (potentially) cause illness at some point down the line, we need to remove that ingredient from cosmetics altogether. But the data is not clear yet.

This begs a bigger, theoretical question. Phthalates and parabens haven't been proven to be medical dangers, but they haven't been disproved either. In fact, some animal data suggests that there is good reason to continue studying these chemicals because they don't appear to be completely benign. As endocrine disruptors that seem to act predominantly on male hormones, phthalates may play a role in the formation of testicular tumors or fertility issues in men. Since parabens show up in breast tumors, it is possible that they have some role in the evolution of the cancer. Just because there is no clear evidence demonstrating that phthalates and parabens in cosmetics actually cause cancer doesn't mean that exposure to a potential carcinogen or endocrine disruptor is okay.

So if you can avoid phthalates and parabens, do so. When it comes to cosmetics, if a plasticizer or a preservative can be left out and the product still accomplishes what you need it to, that's likely the better choice. Your child won't be any cleaner with the phthalate-containing soap or shampoo—it is reasonable to buy phthalate free.

Now, on the flipside, this doesn't mean I am endorsing "all-natural" products, because you don't necessarily get something safe. "Natural" may mean grown in the ground, but that doesn't equal healthy. It may imply "pesticide free" and "**organic**," but unless those words are used, it isn't necessarily the case. While the word "natural" may suggest that they're good for you, natural products aren't always well studied.

There are two problems with trying to avoid phthalates and parabens. First, chemicals in cosmetic products aren't always listed in the ingredients. If the phthalates are part of the fragrance used in a lotion or part of what gives a shampoo its "trade secret," those ingredients don't need to be disclosed.

The second problem is that we can pick on phthalates and parabens today, but there will inevitably just be another additive to pick on tomorrow. Given the way our generation approaches parenting and the way our world disseminates information, there will always be another scare, another toxin to worry about, and another poison threatening our children. As parents, we will have to wade through that hype and decide whether it is legitimate.

In the meantime, if you are looking to avoid phthalates and parabens, get into the habit of reading labels. The Environmental Working Group Web site is a good resource for phthalate information and provides lists of phthalate-free cosmetics. Even though pesticide-free or organic products are not automatically free of phthalates, they are likely to have fewer of them. Meanwhile, the majority of cosmetics manufacturers have chosen to remove parabens from their products. Consumers don't seem to mind the shorter shelf life of their cosmetics when the trade-off is a less chemically laden product.

WHAT'S IN MY HOUSE?

I still haven't gotten into the habit of checking every label for phthalates and parabens. When I do, I go for the plasticizer- and preservative-free brands. But frankly, even I can't always tell if they are in there. Of all the things there are to worry about in this world, the chemicals added to cosmetics are low on my list.

Chapter 18

Deodorant and Antiperspirant

WHAT IS THE QUESTION?

Over the past ten years, one e-mail has made the rounds several times. It comes with a variety of subject headings like "Send this to every woman you love" or "Breast Cancer Prevention." I have received it at least three or four times. The e-mail states clearly and unequivocally that antiperspirants cause breast cancer. Here's the real news flash: they don't.

If the connection between antiperspirants and cancer were really true, its impact would be *huge*. Think of the millions and millions of women all over the world who use antiperspirant. It is true that breast cancer is the second most common type of cancer (175,000 new cases per year) among women and the second most common cause of female cancer deaths (43,300 per year). One would think that if antiperspirant were the cause, the numbers would be even higher.

Several scientists have studied the issue. While there is absolutely no data proving that antiperspirants cause cancer, the research has opened a Pandora's box of other antiperspirant- and deodorant-related health issues.

What do we really have to worry about here? Are the chemicals and heavy metals in these products dangerous? If you are going to use an antiperspirant or a deodorant, which one is safer? How about for the blossoming preteen or teen in your home? Do you need to worry if your child uses one of these products?

WHAT IS THE DATA?

We think of sweat as stinky. But really, human sweat is odorless. That is, until it comes into contact with bacteria.

We all have bacteria on our skin—up to one hundred thousand per square centimeter. The specific bacteria that make sweat particularly pungent thrive in hot, humid, and acidic places like the armpits, feet, and scalp. In these spots, sweat glands secrete extra moisture in an effort to keep things cool. Bacteria flourish in the damp environment, consuming your sweat and all the while releasing a chemical called 3-methyl-2-hexenoic acid: the chemical that makes sweat smell.

Some people don't have stinky armpits. Either those people don't sweat enough to feed the bacteria on their skin or the bacteria living on their skin aren't in sufficiently high numbers to generate the smell. Other people simply cannot smell their own odor. Those individuals are so used to it that their brain doesn't register it as pungent. But for most of us, after exercise or on a hot day, we sweat and it doesn't smell pretty.

Deodorant and antiperspirant are designed to minimize this social faux pas. Deodorant does what its name says: it deodorizes. When you wear it, you still perspire but antiseptic ingredients in the deodorant kill the bacteria where you have applied it so your sweat doesn't smell. Showers and baths actually work similarly—you literally wash the bacteria off. But no sooner are you dry then those same bacteria repopulate the skin and grow right back. Many deodorants are alcohol based because alcohol is effective at getting rid of bacteria. But a few hours later, when the alcohol is gone, the bacteria return. Deodorants are often made with longer-lasting antimicrobials like triclosan. The **FDA** classifies deodorants as cosmetics.

Antiperspirants are completely different. These chemicals actually stop the sweating in the first place. Antiperspirants work by temporarily blocking the pores (technically ducts of the sweat gland) that allow perspiration. If you don't sweat, you won't smell. The active ingredients used to plug pores are almost always aluminum-based compounds.[1] The effectiveness of an antiperspirant varies from person to person—it really depends on how many sweat glands are successfully blocked. Unlike deodorant, the FDA considers antiperspirant a drug.

Both deodorant and antiperspirant have been on the receiving end of bad press. But antiperspirant has had it much worse. As mentioned, starting in the late 1990s a widely circulated e-mail claimed that aluminum in antiperspirants causes breast cancer. The e-mail caught on and has proven to have great longevity (it continues to pop up in my inbox every year or two). Basically, the note lists a variety of reasons why the "chemicals" (in some versions this word is replaced with "aluminum") in antiperspirants are downright dangerous.

The first contention in the e-mail is that the leading cause of breast cancer is the use of antiperspirant. The reader is warned that deodorant is safe but antiperspirant is not, and most of the products sold are a combination of the two. Beware! The author of the e-mail figured out that a simple blanket statement wouldn't suffice, so what follows is a short, semi-scientific explanation about how this could be true. It is written with enough confidence and medical jargon to sound convincing, but in the end almost nothing in the e-mail is factual.

Here's the e-mail's main point: because antiperspirants stop the process of sweating, they interfere with the body's natural mechanism for getting rid of toxins. The lymph nodes in the armpit are responsible for clearing out the waste of the breast tissue (this is true). According to the e-mail, however, if your armpits don't sweat, then toxins accumulate in the lymph nodes under the arm (this is not).

Lymph nodes are in charge of producing immune cells that help to fight infections and removing toxins and cellular debris from the body. The nodes are located in clusters all over the body, connected to each other by tubes that look quite a bit like veins. The lymph system is essentially a waste waterway, the body's own sewage system. The nodes don't rely on sweating to clear themselves of the debris they collect. In fact, sweating plays no role here. The lymph system uses the liver and kidneys to get rid of its identified waste. Whether or not the lymph system does its job effectively has nothing to do with the presence or absence of antiperspirant.

The e-mail then goes on to say that the majority of breast cancers occur in the upper outer quadrant of the breast because this section is nearest the armpit and its lymph nodes. The upper outer quadrant is an anatomical term. By drawing an imaginary line vertically through the middle of the breast and another through the middle of the breast horizontally, the breast can be divided into four quadrants. It stands to reason that the part

---

THE INFAMOUS E-MAIL
(TAKEN FROM URBANLEGENDS.COM)

Breast Cancer Prevention—Not Just for Women

Men don't forget to tell mom, cousins, etc. Deodorants (non-anti-perspirants) are very hard to find but there are a few out there.

I just got information from a health seminar that I would like to share. The leading cause of breast cancer is the use of anti-perspirant.

Yes, ANTI-PERSPIRANT. Most of the products out there are an anti-perspirant/deodorant combination so go home and check your labels.

Deodorant is fine, anti-perspirant is not. Here's why:

The human body has a few areas that it uses to purge toxins; behind the knees, behind the ears, groin area, and armpits. The toxins are purged in the form of perspiration.

Anti-perspirant, as the name clearly indicates, prevents you from perspiring, thereby inhibiting the body from purging toxins from below the armpits. These toxins do not just magically disappear. Instead, the body deposits them in the lymph nodes below the arms since it cannot sweat them out.

This causes a high concentration of toxins and leads to cell mutations: a.k.a. CANCER.

Nearly all breast cancer tumors occur in the upper outside quadrant of the breast area. This is precisely where the lymph nodes are located.

Additionally, men are less likely (but not completely exempt) to develop breast cancer prompted by anti-perspirant usage because most of the anti-perspirant product is caught in their hair and is not directly applied to the skin. Women who apply anti-perspirant right after shaving increase the risk further because shaving causes almost imperceptible nicks in the skin which give the chemicals entrance into the body from the armpit area.

PLEASE pass this along to anyone you care about. Breast Cancer is becoming frighteningly common. This awareness may save lives. If you are skeptical about these findings, I urge you to do some research for yourself. You will arrive at the same conclusions, I assure you.

Thank you.

---

of the breast closest to the armpit (the "upper outer quadrant") is exposed to the most antiperspirant—which is a fair point. As a reader, you are supposed to infer that since more cancers occur in the upper outer quadrant and antiperspirant is applied near the upper outer quadrant, the two are connected—this may be logical, but it is not correct.

Although it is true that nearly half of all breast cancers are in the upper outer quadrant, it is also true that about half of the breast tissue resides in that area. Just because antiperspirant is applied in that neighborhood does not mean antiperspirant causes the tumors found there. The percentage of tumors in that region seems to be pretty proportionate to the percentage of breast tissue.

While this correlation between breast-tissue distribution and cancer locale is quite accurate (and reassuring), emerging data suggests that the number of breast cancers in the upper outer quadrant has steadily increased over the past few years (Darbre 2005). Maybe there could be something at play here after all. Maybe the proportion of tumors found in this quadrant is going to surpass the relative proportion of breast tissue, fueling the argument that something is triggering tumors in this spot. But no one has made a scientific link with antiperspirant.

The closest anyone has gotten is in a 2007 study out of the United Kingdom that analyzed the aluminum content in the breast tissue of breast cancer patients (Exley 2007). In this study, the aluminum content was significantly higher in the upper outer quadrant. But still, there is no causal link to cancer. Also, it's important to note that only women with breast cancer were studied; no breast tissue from a cancer-free woman was included. Even the scientists who orchestrated this study say that it is possible there is more aluminum in the upper-outer-quadrant tissue simply because it is closer to the armpit; those researchers acknowledge that it doesn't mean aluminum caused the cancers. In medicine, we would describe this as: true, true, and unrelated.

Next, the e-mail targets shaving. Women who shave their underarm hair often cut the skin while doing so. They don't realize it because the nicks are tiny, essentially invisible, and don't bleed. But these little perforations in the skin supposedly allow more of the heavy metal in antiperspirant—specifically aluminum—to be absorbed after the antiperspirant is applied.

Studies have looked at this, trying to gauge whether underarm shaving habits and antiperspirant use are associated with breast-cancer formation. There is a correlation. It looks like women who shave their underarm hair more frequently or who have been doing it over a longer period of time are at higher risk for breast cancer (McGrath 2003). But the authors are quick to point out that this is only a correlation; again, causation has not been proved. In other words, while shaving underarm hair and antiperspirant use are associated with breast cancer, it is not at all clear that they cause breast cancer.

In its last, and perhaps most grasping-at-straws argument, the e-mail states that men have a lower risk of breast cancer because they do not shave their underarms. Their hair prevents absorption of the chemicals in antiperspirant. Well, maybe hair could act as a barrier. But that's not why men don't get breast cancer at the same rate as women. Men don't get the disease like women do because they have less breast tissue. About one hundred times less, to be precise. And coincidentally, the rate of male breast cancer is about one hundred times less than the rate of female breast cancer (American Cancer Society).

If the e-mail is so inaccurate, why should I spend so much time on it here? Doesn't that just give it legs? I don't think so, because I believe that anyone—myself included—who gets an e-mail not once, not twice, but several times over several years, with scary subject lines tapping into one of the great fears among women, is going to start to pay attention to it. So when I decided to write a book called *Worry Proof* I had to address this phenomenon.

I sound so resolute—this e-mail is a hoax perpetuating an urban myth!—and yet there is another side. Physicians and scientists have long been aware of the potential health risks associated with aluminum. This metal is a known neurotoxin and has been implicated in a number of chronic diseases, including Alzheimer's disease and end-stage renal disease. We also know that it is absorbed through skin (Anane 1995).

But still, even though aluminum can penetrate our skin and can be toxic to nerves, no study published to date proves that aluminum causes cancer. The National Cancer Institute and American Cancer Society have gone to great lengths to dispel rumors that link aluminum and breast cancer.

A new twist on the aluminum controversy is that this heavy metal may have hormonelike effects. Some researchers think that aluminum can act like estrogen in the body. Anything that might imitate the actions of estrogen might also affect estrogen-sensitive tissue like breast tissue. It's also known that breast tumors often grow more rapidly in the presence of estrogen. If it turns out that aluminum does mimic estrogen, then it could play a role in estrogen-sensitive diseases like breast cancer.

This would take us full circle, back to the notion that antiperspirant, with its aluminum compounds, could in fact be a breast-cancer contributor. But this is all speculative—a reasonable theory but one that has not been proven in studies on humans. There are certainly people who think that if antiperspirant could increase their risk of cancer even by a fraction of a percent, it is not worth using. That reasoning is good, but then these folks need to rethink deodorant too. "WHAT?" you might be thinking to yourself. "Deodorant is supposed to be fine." But if you are going to worry about the risk of aluminum, which is so far unproven and almost certainly small when considered in the bigger picture, you are probably wired to worry about the **parabens** in deodorant too. So this is when I feel compelled to mention that I don't worry about either one.

Parabens are preservatives. Until recently, they were used in a variety of cosmetics, including many deodorants.[2] They are rapidly being phased out, though, as a reaction to public sentiment and scientific research. If your cosmetics do in fact have parabens, they are often listed on the label. Look for methylparaben, propylparaben, butylparaben, or benzylparaben.

There is data suggesting that parabens act as estrogen mimics in the body. And, as I said earlier, estrogen impersonators may affect estrogen-sensitive tissues like breast tissue. But natural estrogen tends to be much stronger—much more bioactive—than chemicals that mimic estrogen. Parabens have been proven to have weak estrogen-like effects, but natural estrogen is thought to play a much bigger role in breast-cancer development than its weak artificial cousin.

There is also evidence that parabens accumulate in breast tissue. A 2004 study documented parabens in eighteen of twenty cancerous breast-tissue samples (Darbre 2004). The researchers, however, did not show that parabens cause breast tumors, nor did they look for parabens in healthy breast tissue or any other tissues in the body. And the study didn't identify the source of the parabens—were they from deodorant or could they have been from soaps, perfumes, lotions, or even foods or drinks?

Wait a second: these arguments are sounding mighty familiar. Aluminum is accused of accumulating in cancerous breast tissue and mimicking estrogen as well. In fact, a number of chemicals discussed in this book are supposedly hormone imitators or **endocrine disruptors** like **bisphenol A** in plastics, **phthalates** in cosmetics, and **phytoestrogens** in soy. Coincidence? No. In medicine we call this a "common final pathway": all of these bad actors ultimately do the same thing in our body. Endocrine disruptors and hormone mimics are the new fearmongers. Here's the catch: we don't really know what they do to us. Scientists can document that aluminum and parabens look like estrogen; they can show that these chemicals fit into specific receptors on cell membranes; and in some cases they can even demonstrate that these chemicals have real biological effects. But we still have no idea what this means in the long run, whether any of these cause chronic disease or cancer. Do we know which endocrine disruptors to be afraid of? Not really.

What about our kids, especially our tweens and teens? They are strongly encouraged to improve personal hygiene. As a result, they represent a huge consumer niche market for antiperspirant and deodorant manufacturers. But this is also a group with raging hormones. Is it safe to add a potential hormone mimic to their equation?

There is no reason to believe that aluminum or parabens or any of the other "chemicals" in these products affect children or teenagers any more than they affect adults like you and me. I have found no studies examining the long-term effects, but it will take years—if not decades—to actually get an answer. There are studies that suggest that women who use antiperspirant more frequently or started at a younger age may be at increased risk for cancer. But other researchers question these results because the studies are small and look at women who have already been diagnosed with breast cancer. What we need is a study that asks teens today to keep track of antiperspirant and deodorant use on a regular basis, and then to follow these girls as they age.

Some think that all this focus on aluminum (and now parabens) steals attention away from more likely causes of breast cancer. In published articles, Michael Thun of the American Cancer Society urges people to focus on two proven and ubiquitous endocrine disruptors in our society: oral estrogen (in the form of oral contraceptive pills) and obesity. Oral contraceptive pills are low-dose hormones used to stop ovulation. An estimated ten million women in the United States (some say as many as fourteen million) take birth control pills. Obesity increases the body's hormone load because the fat cells produce extra estrogen. Two-thirds of

Americans are overweight or obese. For women in one (or both) of these groups, weight or the Pill is a much bigger risk factor than antiperspirant. Thun and toxicologist Philip Harvey have also been quoted suggesting that even if ingredients in antiperspirants and deodorants are proven to have some role in the formation of breast tumors, oral estrogen and obesity are up to ten thousand times more likely to cause breast cancer.

WHAT IS THE BOTTOM LINE?

Antiperspirant and deodorant are not dangerous. There is no evidence that the aluminum in antiperspirant causes breast cancer, despite urban mythology. Some deodorants contain parabens that have been shown to mimic estrogen, but most do not. Again, there is no data that suggests parabens in deodorant cause breast cancer.

If you are anxious to avoid all potential endocrine disruptors, you will have to learn to read ingredient lists and recognize chemical names. The **NIH** has posted a household products database to help consumers. It is a nice search engine that works by product name or by ingredient type.

The known risk factors for breast cancer are diet, cigarette smoking, alcohol intake, age at first period (called "menarche"), older age, and family history. You cannot control your age, how old you were when you had menarche, or your family history of breast cancer. But you can control everything else on the list. If you want to reduce your own risk, eat a diet low in fat, don't smoke, and limit your alcohol intake.

All this applies to your children too. The risk factors are no different. Neither are the potential implications of aluminum and parabens. In my opinion, teaching your child proper hygiene is a critical part of this equation. If a daily bath or shower doesn't prevent body odor, your child probably needs to wear an antiperspirant or deodorant.

WHAT'S IN MY HOUSE?

I use a combination antiperspirant and deodorant. I will confess that the third or fourth time I got the now-infamous e-mail about the evils of antiperspirant, I switched to a plain deodorant. But once I started researching the issue, I was relieved to find very little data damning antiperspirant, and I quickly switched back.

Chapter 19

Diapers

When it comes to diapering a baby, there are two distinct camps: those who are pro disposable diapers and those who are against them. The folks who oppose disposable diapers invoke a series of arguments, most having to do with the environmental impact of all those dirty Pampers and Huggies: Disposable diapers account for an estimated 2 percent of the garbage generated in the United States, and over time they accumulate in landfills, supposedly decomposing. In addition, in order to make these diapers, hideous toxins are released into our environment. Then there's the argument having to do with a baby's health and safety: some critics of disposables say that the plastics and chemicals used to make ultraconvenient, superabsorbent, minimally bulky diapers pose a direct risk to the child.

Are disposable diapers a threat to our babies' health? Is it that much safer to use cloth diapers?

WHAT IS THE DATA?

Diapers as we know them are about 125 years old. Maria Allen is credited with becoming, in 1887, the first person to mass-produce cloth diapers. Of course, cloth diapers existed well before then, but not as a commodity.

About sixty years later, along came the disposable diaper. Johnson &

Johnson sold the first version in this country in 1949. By the early 1960s there were Pampers, and in the late 1970s emerged Huggies. These two brands have been dueling it out for disposable diaper dominance ever since.

Disposable diapers were made possible because three components had been invented: elastic to make flexible waistbands and later thigh bands; a superabsorbent **polymer** called sodium polyacrylate, which absorbs the water in urine and stool; and resealable tape. Together, these three components made disposable diapers functional.

Initially, the disposable diaper was big and bulky. (This rule applies to the first generation of just about everything: remember the first cell phone?) But as materials have been refined, diapers have become downright sleek. There are three layers in a typical disposable diaper. The outermost layer is usually made of a nonwoven material that allows moisture to move away from the skin. The middle layer is typically made of breathable polyethylene or a nonwoven material and film composite, both designed to hold urine and stool. The third, innermost layer contains cellulose pulp and superabsorbent polymers, made from sodium polyacrylate, to lock in moisture.

Sodium polyacrylate is the most contentious material found in disposable diapers. It is used in the form of crystals (which you can actually feel when you squeeze the middle of the diaper). These crystals absorb two hundred to three hundred times their weight in water, turning into a gel in the process. Sodium polyacrylate was used in tampons until 1985, when reports linking it to toxic shock syndrome resulted in its removal. Some people use the tampon example as a reason to not use disposable diapers—how could a parent use a chemical on their child that was banned for adult use? But it bears stating the obvious here: tampons are inserted *into* the body, and the sodium polyacrylate in diapers is not only outside the body but is three layers removed.

What about the crystals parents often find on their children's skin when taking off a wet diaper? Some parents are extremely concerned that these are sodium polyacrylate crystals that have somehow migrated out of the middle of the diaper and that their child has been directly exposed. Not so. When sodium polyacrylate comes into contact with water, it absorbs it and the combination becomes a gel-like substance. The crystals that parents find on their child's skin actually look like crystals or small flakes of rock candy. These are not from the inside of the diaper but are rather a by-product of dehydrated urine called "urate crystals." These form because the diapers are so absorbent that the material pulls away most of the water contained in urine, leaving urate (a normal component) behind. The dehydrated urate forms pinkish orange crystals.

The final diatribe against sodium polyacrylate is that if a child tears apart a diaper and starts to play with the crystals or actually ingests the crystals, there can be toxic effects. This is true: sodium polyacrylate is not meant to be ingested. But young infants have neither the strength nor the coordination to tear apart a diaper. And older children—rascally toddlers who get themselves into trouble at every turn—are still extremely unlikely to take off a diaper and explore its mechanics. Diapers should be stored out of a child's reach or reconsidered entirely if you think that this is a possibility.

There are concerns about disposable diapers beyond sodium polyacrylate—mostly environmental concerns. Perfectly white, sterile-appearing diapers are bleached in order to achieve that look. Bleaching typically uses chlorine, and in the process dioxins are released into the environment. Dioxins are known carcinogens (cancer-causing chemicals). From a public health perspective, the manufacturing of bleached diapers is a reasonable concern.

The other environmental concern is one of sheer bulk. The average American baby uses 5,000–8,000 diapers between birth and potty training. This translates to 27.4 billion diapers used in this country every year. It is estimated that 95 percent of these are disposable diapers. Dirty disposable diapers account for 3.4 million tons of waste every year (or 2 percent of the garbage generated in the United States). The diapers are often sent to landfills, where they are supposed to decompose over six months. But according to the California Integrated Waste Management Board, "Contrary to popular belief, no diaper—not even biodegradable ones—can break down in an airtight landfill. A landfill is not a composting facility . . . nothing degrades well in a landfill." Studies show that, compared to cloth diapers, disposable diapers produce seven times more solid waste when discarded and three times more waste in the manufacturing process.

The last concern about disposable diapers is a relatively new one. A study published in 2000 demonstrated that the plastic linings of disposable diapers caused the temperature of the scrotum to rise. The authors suggested that because sperm production ceases at higher temperatures, perhaps the common use of disposable diapers over the past thirty years accounts for at least some of the increasing fertility issues in men (Partsch 2000). While I don't want to dismiss this suggestion entirely, because I suppose it is possible that a regulatory mechanism for sperm production is somehow altered during the first couple of years of life, it seems pretty unlikely. Think about it: when men are found to have low sperm counts and they are trying to conceive, they are advised to stop wearing tight underpants, stop riding a bicycle, and to get out of the Jacuzzi. These lifestyle changes lower the temperature of the scrotum and often result in improved sperm count. It is hard to believe that warming the testicles a degree or two in the first two years of life—well before sperm are ever produced—affects a reversible problem seen later in life. But I haven't read a study to the contrary, so I will leave that one relatively unanswered.

With this long list of negatives attributed to the disposable diaper, many people feel pressure to use cloth. The modern cloth diaper is a far cry from what it used to be. For starters, it doesn't use safety pins. Most cloth diapers have an outer shell (made from a variety of materials like cotton or microfiber) that snaps or fastens with Velcro around the baby's body and a piece of material to collect the urine or stool. Actual cloth "diapers"—that is, the cloth that is inserted into the shell—are made from industrial cotton, but they can also be made of wool, hemp, or bamboo.

Cloth diapers generally don't look as pristine as their disposable cousins. Still, some of them contain bleached cottons (so the environmental by-product dioxin is still generated) and others have man-made materials like polyurethane laminate, polyester fleece, or faux suede cloth. If you do enough digging, you will probably find something bad written about each of these materials as well.

The health claims against cloth diapers, though, are not related to what they are made from. Rather, the risks are related to how cloth diapers work. The material that collects urine and stool is not superabsorbent. It's basically a towel. Therefore, it doesn't pull the urine and stool away from the skin. So when a baby wets or soils a cloth diaper, the urine or stool sit right next to the skin. This isn't a problem when a parent changes the diaper quickly, but if a child is in a dirty diaper for a while, the chance of diaper rash is fairly high.[1]

Cloth diapers are also not without their environmental costs. Studies estimate that when a parent uses cloth diapers, the family will use fifty to seventy gallons of water every three days, doing laundry. If the family uses a commercial laundry service, the water consumption is lower. But then again, if you calculate the energy consumption and gas used by the laundry service to pick up and drop off the cloth diapers, there is an added environmental expense. Whether you wash your own cloth diapers or send them out to be cleaned, the environmental burden is nowhere near that of disposable diapers, but it's not zero either.

One can get dizzy trying to choose between disposable and cloth diapers. Ultimately, I think most parents opt for the disposable version not just for convenience but because they really seem to keep their babies bottoms cleaner.

WHAT IS THE BOTTOM LINE?

From a health standpoint, there is very little difference between disposable and cloth diapers. Both are safe. While cloth diapers can result in diaper rash more easily, this problem goes away if the diapers are changed as soon as they are soiled. While disposable diapers contain sodium polyacrylate, it is several layers removed from the body, and there is no data suggesting that it is harmful at that distance.

The real issue is one of environmental impact. Disposable diapers fill landfills; cloth diapers, while much more eco-friendly, still consume water. Some cities have established diaper-recycling facilities to proactively address the landfill burden. Also, some diaper manufacturers have produced "hybrid" diapers that have a reusable exterior shell and a disposable core. The idea is to reduce the burden on landfills while maintaining the convenience of disposability. When choosing a diaper, chlorine-free products are probably better for the environment. But it's important to mention that there are manufacturing considerations far beyond the scope of this book.

So when it comes to whether one type of diaper is more or less dangerous for our children than another, there is no real debate. Neither is toxic, and neither is going to cause serious health issues. This is really one where you can decide without any stress, because both types are fine.

WHAT'S IN MY HOUSE?

I always used disposable diapers. For me, the reason was simply to avoid rashes. But if my kids were babies today, I would probably give the hybrid variety a try. It seems to be a great middle ground, and I do have to admit that we generated a lot of trash in our house when our kids were young.

Chapter 20

Insect Repellents

WHAT IS THE QUESTION?

In the past, insects were simply a bother. Mosquitoes bite. Bees sting. We tolerated them, tried to avoid them, and swatted at them, but we didn't really worry too much about them.

In other parts of the world, though, insects have long been recognized as vectors of infectious diseases. Malaria, Japanese encephalitis, yellow fever, and dengue fever are common examples of illnesses passed directly from insects to humans. In the past two decades, the appearance of West Nile virus and Lyme disease has made insect-borne illnesses a concern stateside. One little bite from the wrong animal can lead to a serious infection.

Insect repellents that can be applied directly to the body come as sprays, lotions, and creams. While these topical treatments are able to prevent bites and therefore avert possible serious illness, many people question their safety. What is more harmful: the risk of infection from an insect bite or the potential health effects of the ingredients that make insect repellents work?

WHAT IS THE DATA?

The most commonly used insect repellent is **deet**.[1] It protects us against bites from mosquitoes, ticks, and biting flies. It was first used by the military during World War II, and a decade later it was adapted for civilian use. Because deet has been used for so many years, there is a lot of data about its safety and its risks.

Scientists don't entirely agree on how deet works, but most believe it basically alters an insect's sense of smell. An insect needs to sense human sweat and breath in order to hone in on its target and know to bite. Presumably, deet blocks the smell receptors in the insect. Deet does not kill insects; rather, it is simply a repellent that takes away the insect's instinct to bite.

Depending on the bottle or brand, you will see different concentrations of deet. You can find deet concentrations as high as 100 percent, though most repellents contain 10-30 percent. The strength of deet determines how *long* the repellent will work on the skin, not how well it will work. A 10 percent deet solution lasts approximately two hours, while a 30 percent solution lasts more than five hours. People who use 100 percent deet can get up to twelve hours of protection, though just about all the experts agree that there is very little reason to ever use a solution with more than 50 percent deet.

The **AAP** has published recommendations about how deet can safely be used on children. Initially, the AAP advised using 10 percent deet or less, and only on children over the age of two months. Recently, it has liberalized its policy, suggesting that up to 30 percent deet is safe for kids. While it is still not recommended for infants under eight weeks, using deet at those levels is considered safe for pregnant women.

What's the worry if this chemical simply blocks an insect's sense of smell? The issue is that deet can affect its human user. Mostly, it causes a local skin reaction with itching, burning, or redness. Now, these reactions can occur anytime anything is applied to the skin—even a moisturizer can cause irritation. But with deet, the likelihood of irritation is higher. Some people are really bothered by the fact that deet can alter clothing, actually changing the consistency of some plastics by basically dissolving them. If deet melts plastics, people wonder, what is it doing to their skin?

But the biggest concern about deet is its association with seizures. Let me put this in perspective. There have been ten reported cases of seizures in children after the application of deet, with the last one reported in 1992. Only ten.[2] Moreover, it is very difficult to automatically blame these on deet because seizures are common in children: 3–5 percent of all children have a seizure at some point. Deet exposure is common too, with an estimated 23–29 percent of all children using deet at some point. Teasing these two factors apart has been almost impossible. To add confusion, in several of the cases other diagnoses (like a viral infection of the brain or Reye's syndrome) were never completely ruled out. So while the **EPA** and the **CDC** have acknowledged an association between deet use and seizures, it remains unclear whether deet actually *causes* seizures. If it does, it seems obvious that this is exceedingly rare. As the EPA notes on its Web site, there are an estimated ninety million deet users in this country, but the risk of seizure among deet users is only one in one hundred million.

Parents are generally more concerned about using deet on their children than on themselves. But deet is no more toxic for kids than for adults. A 2002 study by the American Association of Poison Control Centers showed that children were no more likely than adults to have adverse effects from deet.

Okay, so maybe deet isn't so bad after all. Still, people tend to want to avoid it. There are certainly alternatives to deet. In 2005 the AAP came up with a list of insect repellents considered safe for children. The list includes picaridin, oil of lemon eucalyptus, and soybean oil.

Picaridin is the most effective deet alternative. It appeared on the market in 1998 and has been aggressively studied over the past decade. The results show basically the same safety profile as deet, with no accusations of causing seizures and fewer complaints about skin irritation. Also, picaridin doesn't dissolve plastics the way deet does.

Still, compared to deet—with its sixty-plus-year history of data— picaridin is a newcomer. So far it has not been proven to be effective against tick bites, so it is really only useful against mosquitoes and biting flies. Until it is shown to be effective against ticks, certain diseases, like Lyme disease, are better prevented by using deet. All the safety data on picaridin looks promising, but often it is only time that tells.

Oil of lemon eucalyptus is made from eucalyptus leaves and twigs. The natural oil repels mosquitoes and biting flies. The EPA has licensed a synthetic form of its active ingredient called PMD (short for para-Menthane-3,8-diol). Many consumers like oil of lemon eucalyptus for its name—it is a "natural" product rather than being a chemical or drug.

The CDC has endorsed it as well, suggesting that it can be used to protect against West Nile virus in lieu of deet or picaridin.

Oil of lemon eucalyptus is about half as effective as deet at any given concentration. Therefore, to have the same effectiveness as a 10 percent deet product, you would need to use a 20 percent oil of lemon eucalyptus product. Oil of lemon eucalyptus has been on the market since 2000; like picaridin it is a relative newcomer. The only real adverse effect of oil of lemon eucalyptus is eye irritation. But it hasn't been studied in children under age three, so it's not recommended for kids in that age group.

Other "natural" remedies that seem to work as insect repellents include soybean oil (2 percent), citronella, and lavender. The EPA endorsed soybean oil (2 percent) as a reasonable mosquito repellent, but it only works for approximately ninety-four minutes, so many researchers feel that it doesn't provide sufficient protection against mosquito bites and mosquito-borne illnesses. Everyone agrees that it is less effective than deet, picaridin, and oil of lemon eucalyptus.

Citronella can be rubbed on the body or added into candle wax. It's an effective insect repellent but only for a very short time—about twenty minutes. Lavender has a similar downside, with data showing that it is effective for about half an hour. Neither one of these is studied in children under two, so you shouldn't rub it on their skin. In fact, among children with sensitive skin, lavender is a well-known irritant.

Overall, synthetic repellents such as deet and picaridin are more effective than their natural counterparts—they last longer and work better. The question is: do you really need to use insect repellent? The risk of getting bitten by the wrong bug and actually winding up with a profound illness seems so remote. No one wants to expose kids to more and more chemicals. Can't we just skip the insect repellents altogether? Actually, no—because the diseases carried by insects aren't so uncommon and definitely aren't benign.

The two most serious insect-borne illnesses in this country are West Nile virus and Lyme disease. West Nile virus is passed by mosquito bites. The disease first appeared in the United States in 1999, initially making landfall on the East Coast but rapidly spreading westward. In 2007 there were an estimated 175,000 Americans infected with West Nile virus. West Nile virus causes polio-like symptoms, including paralysis, in one out of every twenty people with the infection. One in one hundred infected people get disease in their central nervous system (such as meningitis and encephalitis). In 2007, 1,227 people had West Nile meningitis or encephalitis, and 117 died of the disease.

Lyme disease is entirely different from West Nile virus. In fact, the two diseases only have two things in common: both are passed by an insect bite (though different insects do the biting), and both can cause illness in the central nervous system.

Lyme is caused by a bacterium called *Borrelia burgdorferi* that is transmitted by tick bite. Data from 2005 documents that, in the United States, for every hundred thousand people there were 7.9 cases of Lyme. In the ten states where Lyme was most prevalent, there were 31.6 cases for every hundred thousand people.

If an infected tick bites a person, the symptoms of Lyme disease progress in three distinct stages. First are the nonspecific flu-like symptoms that appear one to two weeks after the tick bite, including headache, fever, muscle aches, and fatigue. Some people also develop a rash called erythema migrans, which is a red, painless rash that looks a lot like the Target logo (a bull's-eye). If the disease is undiagnosed and untreated, it can progress to the second stage. Here there is joint swelling; neurological symptoms like shooting pains or Bell's palsy, which is a loss of muscle tone on one side of the face; and heart palpitations or other heart rhythm abnormalities. Lyme can also go on to a third, chronic stage where the neurological symptoms become more intense, the heart rhythm disturbances persist, and many people develop arthritis and profound fatigue. There are also some cases of psychiatric diseases associated with chronic Lyme.

While Lyme almost never causes death, it can be associated with profoundly debilitating symptoms. This is tragic given that Lyme is preventable (by avoiding tick bites) and treatable if you do get the disease (with an antibiotic).

WHAT IS THE BOTTOM LINE?

The bottom line is that the potential complications of insect repellents, whether chemical or natural, are preferable to the potential complications of a bite from the wrong bug. Lyme disease and West Nile virus are no longer uncommon. Each disease is associated with a laundry list of symptoms, and West Nile can even cause death. I am not trying to be dramatic, but the worst data on deet doesn't hold a candle to the worst data on West Nile or Lyme.

If you choose to use deet, only apply it to exposed skin. On a hot summer day, this means the arms, legs, and face.[3] Any skin that is already covered with clothing is sufficiently well protected. In order to reduce any risk of deet reaction, the less you use on your body, the better.

Of course, better than applying an insect repellent is to protect yourself with clothing and common sense. Wear long sleeves and long pants when the temperatures permit. On very hot days, though, the risk of overheating, heat stroke, or dehydration is not worth the clothing coverage. Don't go into insect-ridden areas if you don't need to. Mosquitoes are known to buzz around and bite at dawn, dusk, and after nightfall. If you are in a mosquito-infested area, avoid outdoor activities at those times.

Don't use a combination sunscreen–insect repellent. The problem with these products is that the sunscreen needs to be reapplied every one to two hours while the insect repellent doesn't. As a result, most people get either too little sunscreen (because they don't reapply) or too much repellent.

If you live in or visit a tick-infested area, check yourself and your children for ticks each day. According to the CDC, a tick must be attached to a human for at least thirty-six hours before *Borrelia burgdorferi* (the bacterium that causes Lyme disease) can be transmitted. If you do find a tick, pull it off with tweezers and clean the area with a topical antiseptic.

WHAT'S IN MY HOUSE?

LA is not known for ticks or mosquitoes, but when we go places where there are lots of these critters, I protect my kids with deet. There was a West Nile scare in my area a few years ago, and all the local parks had prominently posted signs. It is pretty easy to follow their suggestions— my kids didn't play at dawn or dusk, and we steered clear of parks with standing water since that's where mosquitoes tend to congregate. You don't need a medical degree to figure this out. Basic common sense will go a long way in protecting you and your family from insects.

Chapter 21

Sunscreen

WHAT IS THE QUESTION?

The sun is responsible for daylight, solar energy, photosynthesis, and much of what makes the world thrive. But it is also a known evil. It causes sunburns, aging of the skin, and some skin cancers.

Sunscreen is the antidote—hailed as the protector against burns, wrinkles, and skin disease. But recently sunscreen has come under fire. Does it really protect as well as claimed? Are the ingredients in sunscreen—most of which have long, cumbersome scientific names—dangerous? Isn't some sun exposure good? Which is worse: the sun's rays or the sunscreen we use to protect against them?

WHAT IS THE DATA?

The sun emits energy in two forms: visible light and ultraviolet (**UV**) light. The visible light rises in the morning and sets at night. What you see is what you get. UV, on the other hand, is invisible to the naked eye.

There are two types of UV light emitted by the sun that actually reach us here on earth: UVA and UVB. A suntan generally comes from UVB exposure. These rays are present much more during the summertime and much less in the winter. For this reason sunblock—originally produced for summer use—was designed to protect against UVB. UVB is definitely worth protecting against, because exposure is directly related to skin cancers like basal cell carcinoma and squamous cell carcinoma. The UVB waves cause direct DNA damage, turning some cells cancerous.

UVB is not all bad, though. Among its most important contributions, UVB helps the skin produce vitamin D. Vitamin D, in turn, improves calcium utilization, building strong bones. A deficiency in vitamin D can result in rickets. Vitamin D also boosts mood, and low levels are associated with depression.

Until about twenty years ago, UVA was thought to be the sun's "safe" ray. This was a good thing, because an estimated 90–97 percent of the UV radiation that reaches the earth is UVA. But as it turns out, UVA is not as benign as once thought.

Studies show a correlation between sun exposure and the most deadly skin cancer: melanoma. Exactly how the sun causes this cancer has not been proved, but UVA is the leading suspect. Why? Because there seems to be an increased risk of melanoma in some sunscreen users, and until recently most sunscreens only blocked out UVB. That means that avid sunscreen users were still exposed to lots of UVA. In fact, some people may have increased their sun exposure, spending more time outdoors because they thought their sunscreen provided complete protection, and in doing so they actually increased their UVA exposure. With good UVB protection, the skin doesn't burn as easily. So the natural reminder that it is time to go inside is missing. Currently, the most widely accepted theory is that UVA causes indirect DNA damage by generating free radicals and that ultimately results in melanoma.

UVA is not all bad either. It can be used therapeutically, particularly among patients with skin diseases like psoriasis and vitiligo. The UVA waves penetrate into the deep layers of the skin, and in these particular cases the exposure can be more healing than damaging.

Regardless of the upsides of UV rays, they are still more hurtful than helpful. Skin cancer is the most common form of cancer diagnosed in the United States, and more than 90 percent of all skin cancers are caused by sun exposure.[1] In fact, a person's risk for skin cancer doubles if she has had more than five sunburns. Even one severe blistering sunburn puts a person at increased risk for skin cancer. Other risk factors for skin cancer include light skin color, family or personal history of skin cancer, exposure to sun through work or play, light colored eyes and hair, and a large number of moles (more than fifty).

Among skin cancers, basal cell and squamous cell carcinomas are the most common and both are highly curable. Melanoma ranks third in frequency, but it is the most deadly type. In 2004 almost eight thousand people died of skin cancer. This data represents a dramatic rise in melanoma rates, from 1 in 1,500 people in 1930 to 1 in 250 in 1980 to 1 in 120 in 1987. According to some, the risk of melanoma is now 1 in 87 (Rigel 1996). Said another way, the incidence of melanoma has increased more than 1,800 percent since the 1930s.

Certainly, avoiding sun exposure during peak hours (10:00 a.m. until 2:00 to 4:00 p.m. depending upon where you live) and wearing long sleeves and long pants will help. But what if it's a hundred degrees out and long clothing is too hot? What if your child is at summer camp, playing in the good ole outdoors? Sunscreen is the safest alternative, right?

Probably. But critics argue that because SPF—the currently accepted gold standard for measuring the strength of a given sunscreen—only measures the degree of UVB protection, you really don't know how effective your sunscreen is. UVA does not cause the skin to redden (or burn), so there is no good way for chemists to measure how well it is blocked. Therefore, while you may be using a sunscreen with a very high numeric rating (like SPF 50+), that rating doesn't measure UVA protection at all.

SPF also depends on proper sunscreen use. To achieve the coverage advertised on the sunscreen label for full-body protection, at least one ounce of sunscreen must be applied. That is equal to two tablespoons (or six teaspoons), which is a lot more than most people put on at any one time. Sunscreen also must be reapplied every hour or two and after outdoor activities such as swimming and running. Without following these rules, the protection offered by sunscreen drops dramatically. SFP 15 is supposed to filter 92 percent of UVB rays while SPF 30 is supposed to filter 97 percent. This is rarely the case, though, since most people do not apply enough sunscreen to reach the SPF level indicated for a particular sunscreen.

Critics also complain that sunscreen is laden with chemicals that have their own possible carcinogenic (cancer-causing) effects. Sunscreens contain a long list of additives like oxybenzone, titanium dioxide, zinc oxide, avobenzone, octinoxate, and PABA, to name a few. The data on many of these ingredients is confusing.

In the United States, sunscreen is classified as an over-the-counter drug and therefore is regulated by the **FDA**. As part of the regulatory process, there have been studies on each of the ingredients in sunscreens sold in the States. Although there have been some concerns raised by the data, overall each ingredient has been deemed to be less harmful than the unprotected UV exposure of not using sunscreen at all. Among the most controversial ingredients are:

Oxybenzone. Oxybenzone is a principle ingredient in many sunscreens. It has been accused of disrupting the hormones in our bodies, damaging cells, and when used by pregnant moms causing low birth weight in babies. The jury is still out on many of these allegations. The one substantiated claim is that oxybenzone can cause allergic reactions. This is certainly true. But it is not uncommon for any lotion, ointment, or cream to cause a rash or reaction in some users. It doesn't matter if it is a medicine or a moisturizer, anything you put on your skin can cause irritation.

A more debatable question about oxybenzone is what it does inside the body. Oxybenzone is definitely absorbed into the skin, because it can be found in the urine of most users (Hayden 1997). If sunscreen simply sits on top of the skin and is never absorbed, its ingredients are unlikely to have any biological effects. However, when a chemical waste product is found in the urine, this is evidence that it has been absorbed through the skin, and then it has the potential for health repercussions. Absorption of an ingredient is a big deal. The thing is, no one agrees on what exactly oxybenzone does (if anything) once inside.

Titanium dioxide and zinc oxide. In the United States, a number of agencies study the ingredients in sunscreen. Among the well-respected sources on carcinogens are the International Agency for Research on Cancer (**IARC**), the Food and Drug Administration (FDA), and the ***Report on Carcinogens*** (a published list of known or reasonably anticipated human carcinogens). None of these institutions considers titanium dioxide or zinc oxide carcinogenic. The *Report on Carcinogens* and the IARC do, however, list solar radiation (sun exposure) as a known carcinogen.

A recent study done by Australia's Department of Health and Ageing looked specifically at the safety of titanium dioxide and zinc oxide in sunscreen. Australia is famous for its sun protection efforts: every schoolchild is expected to wear a hat during the day to minimize exposure. The report states that both titanium dioxide and zinc oxide remain on the skin surface in the outer dead layer. They don't penetrate living skin cells. The authors conclude that even if titanium dioxide and zinc oxide were carcinogens (and there is no evidence for this), these substances are not absorbed so there is no impact on the body.

PABA (para-aminobenzoic acid). PABA was once a staple ingredient in most sunscreens, but it is rarely used now. Many sunscreens actually advertise themselves as "PABA-free" on the label. The main problems with PABA had to do with the fact that it required alcohol in order to work. The alcohol stained clothing and was associated with a number of reactions like allergy, rash, itching, and stinging.

INGREDIENTS IN SUNSCREEN		
Drug Name	**Concentration, Percentage**	**Absorbance**
Aminobenzoic acid	Up to 15	UV-B
Avobenzone	2–3	UV-A I
Clinoxate	Up to 3	UV-B
Dioxybenzone	Up to 3	UV-B, UV-A II
Ecamsule*	2	UV-A II
Ensulizole	Up to 4	UV-B
Homosalate	Up to 15	UV-B
Meradimate	Up to 5	UV-A II
Octocrylene	Up to 10	UV-B
Octinoxate	Up to 7.5	UV-B
Octisalate	Up to 5	UV-B
Oxybenzone	Up to 6	UV-B, UV-A II
Padimate O	Up to 8	UV-B
Sulisobenzone	Up to 10	UV-B, UV-A II
Titanium dioxide	2 to 25	Physical
Trolamine salicylate	Up to 12	UV-B
Zinc oxide	2 to 20	Physical

Source: From http://emedicine.medscape.com/article/1119992-overview.
*Only available in United States in patented products.

With these and other manufactured ingredients, the controversy over the chemicals in sunscreens is not unwarranted. There are several studies under way looking at the potential risks of these ingredients. Many people want to find sunscreens without chemical ingredients, but "all-natural" or "herbal" formulations are not necessarily the answer. Many of the ingredients in those products haven't been well studied—in fact, many "natural" ingredients are completely untested—so those sunscreens may not protect against the sun and could pose health risks of their own. Why should we believe that an untested product is safer than something undergoing rigorous study? Somehow consumers have convinced themselves that no information is better than actual data. It is a modern twist on an old adage: ignorance is bliss.

There are the people who are worried about specific ingredients, and then there are others who worry that using sunscreen may lower vitamin D levels in the body. Sun exposure—and specifically UVB exposure—stimulates the body to produce vitamin D. This is the positive side to the sun exposure coin: in the cave days, there certainly wasn't any sunscreen available, so the human body benefited by using the sun's rays in a positive way. To this end, vitamin D is known to improve calcium utilization and strengthen bones, and to boost mood.

Some studies have shown that vitamin D can actually prevent certain forms of cancer. Absorption of UVB is blocked by sunscreen. Some sunscreen critics have used this to turn the argument on its head: sunscreen may actually cause cancer by reducing the amount of vitamin D produced by the body and therefore reducing the body's natural anticancer drug. I think this argument is a pretty big stretch. Research indicates that our bodies produce plenty of vitamin D with just a tiny bit of UVB exposure. If a person gets UVB on the face and back of the hands—areas commonly uncovered (or minimally covered) with sunscreen—for fifteen minutes a day just two or three times a week, there can be enough vitamin D produced to keep the body happy. In addition, many foods, like dairy products and orange juice, are supplemented with vitamin D. And multivitamins typically contain vitamin D. So there are sources of vitamin D other than the sun to help maintain our health.

WHAT IS THE BOTTOM LINE?

Regardless of the controversies over sunscreen, we know that the sun presents its own set of risks. It is estimated that 80 percent of a person's lifetime sun exposure occurs before the age of eighteen. This has to do with play patterns (kids play outside much more than their adult counterparts), inability to self-regulate when overheated (kids will often just keep on going, even when they are very hot), inadequate use of hats and sun-protective clothing (it is the rare child who will keep a hat on, except in Australia), and lack of consequential thinking (a mother concerned with the aging effects of the sun or fearful of sunburn is more likely to take precautions than a child who is only thinking about playing in the moment).

To protect against the risks of sun exposure, use common sense and minimize your—and your child's—sun exposure at peak times. Use protective clothing and hats. When you need to apply sunscreen, use it properly. Depending on a person's size, it takes an ounce of sunscreen or more to adequately cover the entire body. You need only apply it in sun-exposed areas, but be generous and reapply every one to two hours and after swimming. There is no point to using just a little sunscreen—if you do, you run the risk of a burn plus exposure to the ingredients that some question.

Infants under six months really shouldn't be in direct sunlight for more than a few minutes at a time. Their skin has very little melanin, so it can burn easily. Babies also cannot get out of the sun's rays even if they want to because they are immobile.

As for sunscreen ingredients, oxybenzone is absorbed, proven by the fact that it shows up in urine. Few other ingredients hold this distinction. I would limit the use of oxybenzone by opting for a sunscreen without

SPF AND SUNSCREEN APPLICATION TECHNIQUES

SPF 15 sunscreen provides good UVB protection for healthy individuals. By definition, SPF 15 filters out more than 93 percent of UVB radiation, and a SPF 30 product filters out more than 97 percent. But as it turns out, these percentages are based on proper application of the sunscreen. When the FDA tests SPF, the standard is to apply sunscreen at a thickness of 2 mg/cm². Several studies show that in real life application thickness is more like 0.5–1.0 mg/cm². That is one-quarter or one-half the studied and recommended amount. It is clear that at these lower "doses," the effective SPF is significantly reduced.[2]

it until the biological effects are clarified. Keep in mind that if you use opaque sunscreens with titanium dioxide or zinc oxide, they work by reflecting sunlight, so you need to see the white residue on your skin in order to be adequately protected.

WHAT'S IN MY HOUSE?

My son is blond and fair, so it is easy to remember to put on the sunscreen. It is harder to remember with my daughter, who has dark hair and olive skin and—lucky her—will probably never burn. But she still needs sunscreen. So I keep it in the kids' bathroom next to the toothpaste, and my children know that putting it on is a part of the morning routine.

Chapter 22

Toothpaste

WHAT IS THE QUESTION?

Dental hygiene begins when your first tooth appears and should continue for the rest of your life. Brushing your teeth is one of the few things that you do every day—twice a day or more—for years and years and years.

Toothpaste is designed to keep our teeth healthy and strong, to prevent cavities, and to freshen breath. Different combinations of additives and ingredients make your teeth pearly white or your breath minty clean. Go to the supermarket or drugstore and it can be overwhelming to pick one out among the dozens of choices available.

With all these brands and flavors and claims come a series of questions. Are there ingredients in toothpaste that are dangerous for my child? Is fluoride necessary or is it unhealthy? What about the colorings and flavorings in children's toothpaste? There are lots of "natural" alternatives available—are these any better or safer?

WHAT IS THE DATA?

The toothbrush was presumably invented in China in the 1400s. At least, this is the first time that a toothbrush can be documented. But it took four hundred years for the toothbrush to come west, making its debut in Europe in the 1800s. Of course, toothbrushing doesn't necessarily require a toothbrush—there were other ways to do the deed. That said, toothpaste never would have come along if the brush hadn't been invented first.

Initially, people only used water with their toothbrushes. In the late 1800s, they began to put powders on the brush in order to improve cleaning. The powders were meant to be abrasive and so they were made from a variety of substances like chalk, brick, and salt. Around 1900 a paste made from hydrogen peroxide and baking soda became popular. At the same time, the concept of putting the paste into tubes took off. So it was only a little more than a hundred years ago that the first tubes of toothpaste appeared on the market.

Fluoride was added to some toothpastes starting in 1914. This additive was initially criticized by the American Dental Association (ADA). In fact, the ADA did not endorse its inclusion until forty years later. Opinions changed largely because of the research started by Frederick McKay, a Colorado dentist in practice during the first three decades of the 1900s. He recognized that many of his patients had brown stains and bumps on their teeth, but they had fewer cavities than expected. McKay suspected and later demonstrated that this was due to high levels of naturally occurring fluoride in the drinking water.

In the early 1940s, H. Trendley Dean calculated the ideal amount of fluoride in drinking water—enough to prevent cavities but not enough to create stains or mottling of the teeth. Shortly thereafter, the government agreed to study population effects of fluoride by adding it to the public drinking water. In 1945 Grand Rapids, Michigan, became the first U.S. city to have fluoridated water.

Over the next ten years, data clearly supported the fact that fluoride prevents tooth decay. This accumulating data helped turn the tide for fluoride in toothpaste. In the 1950s, the ADA fully endorsed fluoride-containing toothpastes.

There are always naysayers, and this isn't a bad thing. Critics insist that data is checked and rechecked, that the pros outweigh the cons. So no sooner had fluoride been discovered than it was criticized. The claims against it are important to note, but at least for now the benefits of fluoride outweigh the risks. Here is a partial list of what critics accuse fluoride of causing: vomiting, diarrhea, seizures, tremors, respiratory distress, drooling, heart attacks, heart rate abnormalities (particularly slowing the heart rate), salty or soapy taste in mouth, and weakness.

WHO REGULATES FLUORIDATED DRINKING WATER?

From *Morbidity and Mortality Weekly Report*, July 11, 2008:

Since 1945, the U.S. Public Health Service and CDC have tracked the number of persons in the United States receiving fluoridated water. The U.S. Environmental Protection Agency (EPA) does not regulate water fluoridation, and EPA's Safe Drinking Water Information System only tracks fluoride concentrations in water systems with naturally occurring fluoride levels above the established regulatory maximum contaminant level (4.0 ppm). Water fluoridation is managed at the state level, and CDC relies on states to provide data on individual community water systems (e.g., population served, fluoride concentration, and fluoride source).

There are some other concerns too, like that fluorosis, or white staining of normally off-white teeth, is associated with too much fluoride consumption and can actually increase the likelihood of cavities. There is the possibility that ingesting fluoride weakens bones and therefore increases the risk of fracture. Some studies question whether a high dose of fluoride has reproductive effects by acting as an **endocrine disruptor**. There is even some data from China suggesting that fluoride consumption may lower IQ. The newest claim is that fluoride binds aluminum and therefore may be associated with development of Alzheimer's disease.

It is true that if large quantities of fluoride are ingested, the result can be devastating. One fatality was reported to the American Association of Poison Control Centers in 2002, but there have been none since. According to Poison Control Center data published on eMedicine, death can result from ingesting as little as 2 grams of fluoride in an adult or 16 mg/kg in a child. For an average nine-year-old weighing sixty-six pounds (30 kg), this means 480 mg of fluoride could be fatal; for an average one-year-old weighing twenty-two pounds (10 kg), the fatal dose is 160 mg, which is about the amount that is in a standard-sized toothpaste tube.[1]

But fluoride can have toxic effects on a child at much lower doses. Reports of fluoride toxicity are not uncommon, and they have been reported with as little as 3–5 mg/kg (about a quarter of the fatal dose). In 2004 alone, there were 21,890 reports of fluoride ingestions among children under six. Only 440 were treated in emergency rooms, and none resulted in fatalities.

A more common complaint about fluoride—one that affects many more people than fluoride toxicity—is staining of the teeth called fluorosis. Our teeth are not perfectly white—they are slightly off-white. People with fluorosis have spots of brighter white discoloration on their teeth. In extreme cases, the teeth can look pitted or have brown marks. Developing teeth are particularly susceptible to fluorosis, and this actually drives the warnings about how much fluoride a child should use. Toothpaste tubes are often labeled with cautionary words: parents are advised to use no more than a pea-sized amount of toothpaste for children under six and no fluoride for children under two. Turns out, these warnings are a bit overly aggressive. But the issue is that toothpaste tastes downright good—who hasn't watched their child suck the toothpaste off a toothbrush like it is dessert? Brushing with a small amount of fluoridated toothpaste will not cause fluorosis, but eating large amounts might. Toothpaste "dosing" is conservative for good reason. For the youngest brushers, spitting is nearly impossible. This is why they are advised to use fluoride-free pastes.

Fluoride receives the majority of the negative attention directed at toothpaste, but it is not the only villain in the story. Other additives, such as sodium lauryl sulfate (SLS), triclosan, and zinc chloride have also had their fair share of bad press. SLS is a foaming agent that also dries mucus membranes. It can account for some forms of mouth ulcers. Triclosan and zinc chloride are antibacterial agents that help to prevent gingivitis. They have been accused of links to cancer, developmental delays, liver toxicity, and other ailments, though these claims are unproven and considered quite controversial, even fringe. Add to this group a list of sweeteners and artificial colors (not to mention whatever ingredients are necessary to make kid's toothpaste sparkle), and there is a lot to complain about.

Despite all the claims, there is not much data suggesting that any of these ingredients—especially in small amounts and with proper spitting—is very dangerous. There was one recent health emergency related to toothpaste, though. In 2007 millions of tubes of toothpaste imported from China were recalled after the **FDA** warned that they might contain a chemical called diethylene glycol that is found in antifreeze. The diethylene glycol was apparently used as a sweetener.

WHAT IS THE BOTTOM LINE?

Dental hygiene has come a long way, especially over the past century. As the average life expectancy continues to increase, we need to keep our teeth healthy for a longer period of time. This means avoiding cavities and gingivitis. Dental caries (i.e., cavities) are one of the most common diseases worldwide; fluoride is the most effective preventive public health measure for it.

Using fluoridated toothpaste is not dangerous, and neither is using toothpaste with SLS, triclosan, or zinc chloride. Even the colored, sweetened kids' toothpastes that seem so counterintuitive (how can something so artificial looking and tasting actually clean my kids' teeth?) are not bad. But their safety and effectiveness rely on the fact that we spit them out. Toothpastes are not meant to be eaten, which is why recommendations and formulations vary between children and adults.

There are pastes with fewer additives that work just as well as the big brand names. Tom's of Maine is the only company currently offering fluoride-containing toothpaste that is both "natural" and ADA endorsed. If you are looking for toothpaste that will keep your teeth healthy with as few chemicals as possible, Tom's is a good choice.

WHAT'S IN MY HOUSE?

I have never bought a kids' toothpaste for my children. They have occasionally appeared in my house, after much begging, when someone else takes them to the supermarket, but even my five-year-old and my three-year-old question the artificial taste. As soon as my kids' dentist gave me the go-ahead—when each was about two years old—I started putting the thinnest imaginable film of regular adult toothpaste on their brushes. Because of that, my kids have learned to use toothpaste sparingly and to spit well. As a result, I am really not concerned that any of the ingredients will be ingested, absorbed, or ultimately harmful.

Part V

Medicines

Chapter 23

Antibiotics

WHAT IS THE QUESTION?

The discovery of antibiotics marked one of the major turning points in medical history. For the first time, medicines existed that could fight infections head-on. Diseases that could kill suddenly became fully treatable. Antibiotics were silver bullets.

Of course, antibiotics weren't silver bullets for *everything*. First of all, antibiotics only fight infections. These drugs don't treat heart disease, replace the need for many kinds of surgery, or help with swelling or joint pain or headaches. Antibiotics just treat infections. Second, antibiotics only treat one kind of infection: bacterial. In fact, each specific antibiotic only fights a few particular bacteria. So although the introduction of antibiotics represented a massive shift in one part of medicine, their utility has always been limited.

Nonetheless, the more accessible antibiotics became, the more often these drugs were used. A hundred and fifty years ago, a doctor relied on his clinical skills to make a diagnosis—there was no laboratory to run blood tests and no X-ray, CT, or MRI scanners to take pictures of the inside of the body. With the advent of antibiotics, doctors added an extra weapon to their artillery. Just in case an illness was caused by infection, the infection could (maybe) be treated by having the patient start an antibiotic.

From there it became a slippery slope to the abuse of antibiotics. When given "just in case," antibiotics became grossly overused. Almost immediately, bacteria evolved strategies to live through the antibiotics, so higher doses were required and newer formulations were needed. Over time, the race between drug and bug progressed to where we are now: there are bacteria resistant to almost all antibiotics that exist, and we cannot come up with new antibiotics fast enough. With some infections, we have essentially gone back to square one.

So given this state of affairs, do antibiotics even work anymore? Are they safe? If you give them to your child (or take them yourself), will you put your child (or yourself) at risk of becoming resistant?

WHAT IS THE DATA?

In order to understand the controversy surrounding antibiotic use today, it is important to appreciate the history of antibiotics. This is a pretty short story, lasting hardly more than a hundred years.[1]

In the late 1880s, the **germ theory** of disease became widely accepted. This was revolutionary, not unlike when people embraced the idea that the world was round. This theory, the brainchild of Louis Pasteur, basically suggested that certain germs caused specific illnesses. Now germs come in many flavors, and this knowledge unfolded over the following decades. There are bacteria, viruses, and yeast (also called fungi), among others. Even today scientists continue to discover new types of germs, such as the prions that cause mad cow disease. But we are jumping ahead a bit.

Once the germ theory was accepted, scientists could focus on developing medications that kill germs. In 1929, quite by accident, Sir Alexander Fleming discovered that a certain mold produced a chemical that stopped the growth of a bacterium called *Staphylococcus aureus*. He named the mold *Penicillium* and was credited with discovering the first antibiotic.[2]

The substance produced by *Penicillium* killed bacteria, and in doing so it was given the name "antibiotic." "Anti" means against and *biotikos* is a Greek word meaning "fit for life." So "antibiotic" refers to a substance produced by a living organism that interferes with the growth of another living organism. These days, lots of drugs prescribed by doctors are called antibiotics, but most are synthetic, manufactured in a lab and not by a living organism.

Antibiotics work in one of two ways: they either kill (called "bactericidal") or stop the growth of (called "bacteriostatic") bacteria. These medicines do not affect viruses or yeast. While there are many medicines available today that do treat viral or fungal infections, they are technically not antibiotics but rather are antivirals and antifungals.

Now back to the history of antibiotics. World War II catapulted antibiotics to fame. Penicillin was used to treat infections among soldiers, who provided an excellent (and large) study group. Penicillin worked so well that it entered mass-production starting in the early 1940s. Over four decades, multiple new classes of antibiotics were discovered, generating a surplus of new types of medicines to fight infections. These included the sulfonamides (which were actually discovered in the 1930s just after penicillin); aminoglycosides, tetracyclines, and macrolides (in the 1940s and '50s); and quinolones and trimethoprim (in the 1960s). This period also saw the discovery of antifungals (nystatin in the 1950s) and antivirals (idoxuridine and methisazone in the 1960s).

So over a span of nearly forty years, bacterial infections went from largely untreatable to utterly beatable. Antibiotic research and discovery was progressing at breakneck speed. By the late 1960s, many formerly deadly infectious diseases—like strep throat, which, if untreated, can cause rheumatic fever—had almost disappeared from industrialized nations. As a result, people teetered on the brink of arrogance about this medical victory. In 1969 U.S. surgeon general William H. Stewart announced: "The time has come to close the book on infectious disease" (Nelson 2003).

Why, then, are infections still around? With such a success story, shouldn't most bacterial infections be things of the past? The answer lies in bacteria's ability to develop resistance to antibiotics. Science had won the battle but not the war. There was actually evidence of antibiotic resistance from day one, but it was initially uncommon. Over time, the discovery and development of antibiotics has become a game of survival of the fittest for the bacteria, and a few bugs emerge victorious at the end of each round. Here's how it works.

Occasionally, a small group of bacteria colonize (that is, attach to and grow on) a part of the body. It could be your inner ear, your sinus, your urinary tract, or a part of your skin that you recently scraped. It doesn't much matter. Unencumbered, the bacteria multiply steadily.

Now, if you are prescribed an antibiotic—and if it is the appropriate antibiotic for the specific bacteria—the antibiotic will either stop the bacteria's ability to multiply or it will actually kill the bacteria. Either way, if the antibiotic is 100 percent effective, all the bacteria will eventually disappear. However, what if the antibiotic is only 90 percent effective? What if one out of every ten bacteria exposed to the antibiotic survives and goes on to multiply? Or even one out of every hundred or thousand? Then not only do you still have to contend with the consequences of having bacteria around (often with symptoms like fever or pain), but the bacteria that survived are the ones able to withstand the antibiotic. These remaining bacteria had something in their genome, or mutated and accidentally created something in their genome, which allowed them to survive despite being bathed in the medicine. These are the so-called resistant bacteria. Now their offspring—which share the same DNA with the genetic ability to live and multiply despite a specific antibiotic—repopulate the colony. The new colony is filled with resistant bugs.

When you think about it, the evolution of resistance makes sense. The few bacteria that can live through a certain antibiotic at a certain dose then reproduce and make more bacteria that can live through that same antibiotic at that same dose. If your doctor prescribes a higher dose or chooses a new antibiotic altogether, and if a few bacteria still survive and multiply, these bacteria are even more resistant because they have now survived two separate insults.

Realistically, how big of a problem is this? These days it is shockingly massive. Consider hospital patients, who reside in buildings filled with other sick people. Hospitals represent some of the largest collections of infection imaginable. It is estimated that approximately two million people acquire infections while being in hospitals every year. These are infections those patients did not have when they first walked in the door. An estimated twenty thousand (1 percent) of them die of these infections. And when biologists and physicians look at the bugs causing these infections, they find that a full 70 percent are resistant to at least one drug (and some are resistant to all).

What about outside the hospital? In the world of infectious diseases, infections acquired outside the hospital are called "community acquired" (as opposed to those acquired inside the hospital, called "nosocomial").

The first time a community-acquired bacteria was noted to be resistant to antibiotics was more than fifty years ago, in the 1953 shigella outbreak in Japan. One particular strain of shigella was resistant to multiple drugs. These days, it is not uncommon to find resistant community-acquired infections, especially among certain bugs, like staphylococcus and pneumococcus. It is so routine, in fact, that a quarter of community-acquired bacterial pneumonias are resistant to penicillin. Multiple drug resistance among community-acquired infections is equally commonplace these days, with 25 percent of pneumonias resistant to more than one antibiotic.

Two community-acquired drug-resistant bacteria have become so ubiquitous that they are known even among the lay public by their acronyms. **MRSA** is methicillin-resistant *Staphylococcus aureus*, and VRE is vancomycin-resistant enterococcus. Their rise to infamy is scary and bears noting.

Basically, MRSA and VRE developed mechanisms to survive antibiotic treatments. The more doctors overused antibiotics, the more resistant bacteria were selected to grow and survive. These bugs spread easily. Today they are downright commonplace. According to the CDC, invasive MRSA doubled in the first five years of this century. What's more, 15 percent of these MRSA infections (approximately 14,000) occurred in people with none of the known risks factors: they were not hospitalized, not immunocompromised, not sick. More people in the United States now die from MRSA infection than from AIDS. In 2005 MRSA caused an estimated 94,000 life-threatening infections and 18,650 deaths; that same year, approximately 16,000 people in the United States died from AIDS (Boyles 2007).

How does something like this happen? How did we find ourselves in a crisis of antibiotic resistance? There are a few answers.

A parent calls the pediatrician and describes an illness. The child has a cough, runny nose, and fever. The parent cannot bring the child to the doctor for any number of reasons—work, travel logistics, there are other kids in the home, whatever—and the doctor advises the parent what to do: make sure the child rests, give a fever reducer when needed, maybe put a humidifier in the bedroom, and so on. Two days later, the same parent calls back. The child is still sick and, no, the parent and child cannot come in. Now the parent has missed two days of work, and life is getting complicated. Won't the doctor please prescribe an antibiotic? Okay, says the doctor, fine. Within twenty-four hours the child is better. The parent feels victorious—it must have been the antibiotic. Good thing I called the doctor back. But the pediatrician knows that most viruses resolve on their own in three to seven days, and it is just as likely that the child got better because it was time to get better. Without examining the child, the doctor would never know the true cause. Now, multiply this scenario by a million (or several million if you want to consider all the adults who call their doctors when they get sick themselves), and you have the makings of an antibiotic disaster. This has been one major contributor to the evolution of antibiotic resistance.

Another major contributor is lack of compliance. There are people who really need antibiotics—people who are diagnosed with bacterial infections that will not get better on their own. The dose of the medicine, number of doses per day, and total number of days the medicine needs to be taken are not arbitrary. There are guidelines for doctors about how to treat specific infections appropriately in an effort to curb ongoing resistance. But many patients go home with a prescription to be taken three times per day for ten days and only take the medicine a couple of times a day or they stop after a week because they feel better. Those people foreshortened their antibiotic course. As a result, these same patients often find themselves sick again a week or two later because they have not completely eradicated the infection. A patient may have killed off 99 percent of the bacteria, but the 1 percent that remained has now multiplied, so she is sick again. Because these patients have exposed the initial bacteria to antibiotics, there is a good chance that at least some of the surviving bugs are resistant.[3]

The most surprising reason for resistance is that we are exposed to antibiotics, without even realizing it, through agriculture. It turns out that animals consume more than half of the antibiotics ingested in this country. In 1998 alone, humans filled eighty million antibiotic prescriptions. If you put all those drugs on a scale and weighed them, they would equal 12,500 tons (Todar 2008). But agricultural practices used 18,000 tons that year. Agriculture—the use of antibiotics in feed, in plant material, and to proactively ward off infections so that animals can grow big and strong—accounted for 60 percent of antibiotic use.

Our consumption of antibiotics in food comes mostly from eating meat (every kind, including red meat, chicken, pork, and turkey) as well as dairy. But vegetables and grains contribute to antibiotic resistance too. Yes, vegetables. Antibiotic-resistant genes find their way into our produce supply through genetic modification. When a crop is genetically modified, antibiotic-resistant genes are often used as "markers" to identify whether or not the genes were effectively integrated. The genes are actually inserted into the crop's DNA.[4] From there, the genes are ingested, not just by humans, but also by microbes in the environment. Pretty heady stuff: antibiotics and mechanisms for antibiotic resistance are all around us, all the time.

Ultimately, we are going to have to find a way to get resistance under control. This is because neither physicians nor drug developers can keep up with the evolution of bacteria—these organisms are developing strategies for surviving antibiotics faster than we can develop new drugs. To this end, doctors are trying hard to use fewer and fewer antibiotics in order to slow the progression of resistance. So they are prescribing antibiotics less often. But this reduces the drugs' profitability for the drug companies that manufacture them. The decrease in profit makes it less likely that a company will invest more money to develop new antibiotics that will, in turn, be prescribed even more sparingly.

This situation is made worse because, while it used to be that an antibiotic could be used for years, with the rapid emergence of resistance an antibiotic that is very effective one year may become almost useless the next. This means that the investment a drug company makes to bring an antibiotic to market is likely to return an even lower yield because the medicine is apt to have a very short shelf life.

In addition, the business of developing new antibiotics has become increasingly cumbersome and expensive. In the mid-1980s, sixteen antibiotics were approved over a five-year span; compare this to the past five years, when only five were approved. It takes eight to ten years and costs anywhere from $800 million to $1.7 billion to bring an antibiotic to market. While antibiotics are still profitable—raking in an estimated $23 billion in worldwide sales—it is much more profitable to design drugs that people must take every day indefinitely, like cholesterol-lowering agents or blood pressure medications.[5]

WHAT IS THE BOTTOM LINE?

Antibiotics (along with their preventive counterpart, vaccines) are credited with adding an average of twenty years to the life spans of people living in developed countries. These are not evil medications. When used appropriately, antibiotics can be miracle cures.

But these wonder drugs haven't always been used appropriately. Doctors and consumers alike fell in love with antibiotics and, for a time, overused them. Now we find ourselves wrestling with resistance.

Doctors have wised up to antibiotic overuse. Young doctors are trained to use them sparingly; hospitals have protocols for which antibiotics can be used, and when, in order to minimize the emergence of newly resistant strains. But patients are still a step behind. While there are some people who avoid antibiotics to a fault (which isn't smart either, since there are times when the drugs are warranted and even critical), there are others who beg for, or even demand, them. Patients should learn to rely on their doctors to determine when an antibiotic is necessary and appropriate. It is also important to follow the prescribed course of an antibiotic and not to adjust it according to one's own perceived medical expertise and one's current symptoms or lack thereof.

Meanwhile, we should all be advocating the removal of antibiotics from our food supply. Again, this is partly the fault of the consumer: we have become accustomed to bigger pieces of meat and perfect-looking produce. But when we consider why our food looks this way, we find that antibiotics play a significant role. It is critical that consumers begin to adjust their demands for bigger, shinier, meatier, glossier, juicier foods. Shopping at Whole Foods for your own family doesn't solve the bigger problem of antibiotics in our national food supply. Even if you are moderating your own family's grocery list, your family still gets lots of exposure through restaurant foods and eating at friends' homes. It is of utmost importance to take steps to stop the overuse of antibiotics in our meats and dairy. Perhaps that means writing a letter to Congress while also using the power of the purse and no longer buying the adulterated-looking items at the store.

The pharmaceutical companies, meanwhile, need to carry on their quest to look for new antibiotics. There should be some incentive for doing so, whether it involves public funding or an acceptance that for the greater good development of these drugs may wind up being analogous to a lawyer's pro bono work. Perhaps the profit margins for pharmaceutical companies are typically so high that they need to consider making less on this specific class of drugs.

WHAT'S IN MY HOUSE?

As a mother, I really do practice what I preach. I won't give my kids antibiotics unnecessarily, but each has needed them in the past—and took the full course. As a doctor, I have had experiences when I had to refuse prescribing antibiotics—almost to the point of a fight. If the illness is clearly viral, antibiotics will do no good. I have also had the opposite experience, where the child was quite ill with a bacterial infection and I literally had to beg a parent to give the antibiotics I prescribed. I believe that if people better understood when and why antibiotics need to be used, they might ultimately become better advocates for their own health and the health of their children.

Chapter 24

Cough and Cold Medicines

WHAT IS THE QUESTION?

The cold medicine aisle in the supermarket or drugstore is lined with drugs promising to help coughing, congestion, runny noses, allergic symptoms, fever, and any given combination of cold and flu symptoms. There are hundreds of variations on the same theme carried under dozens of brand names. It is overwhelming, even for a mother who is a pediatrician!

When patients call me for advice about managing a child's run-of-the-mill cold, I generally prescribe the same thing over and over: lots of fluids, plenty of rest, and a fever reducer if the temperature is high and your child is uncomfortable. Sometimes an over-the-counter (**OTC**) medication will help an older child, but don't give your young child any cough and cold medicines because they won't work. The virus will run its course.

Turns out, the **FDA** agrees with this advice. In 2007 it took it one step further, saying, not only are cold medicines ineffective, but they can actually be dangerous. The government recommended voluntary relabeling so that the medicines were not given to children under two. One year later, the instructions were changed again, telling parents, "do not use in children under four."

What is so dangerous about OTC cough and cold medications? Should we be freaked out that we have used them for years and years?

WHAT IS THE DATA?

Cough and cold medicines are a huge industry, with more than eight hundred different varieties on the market. It is estimated that every year ninety-five million pediatric cough and cold medicine packs are sold (Zwillich 2008).

This number seems unbelievably high until you think about it for a minute: your child is sick and you will do anything you can to help him feel better. You go to the drugstore and wander down the aisle, trying to figure out which bottle to choose. There's "cough and cold," "sinus symptom formula," "cough plus fever reducer," "congestion relief," and on and on. There are liquids, melt-away strips, and pills. There are flavored (berry, cherry, orange, orange-berry-cherry) and color-free, dye-free, taste-free formulations. Here you are, among the hundreds of choices. Of course you are going to pick up at least two (maybe three or should you choose four?) just in case your child doesn't like the taste. You don't want to have to come back at two in the morning. Now multiply this scenario by millions of parents.

You get home and try the first medicine. He's so sick he actually takes it (which is a small miracle in itself). But an hour later he's no better. He still feels warm; he's still congested. It's now getting late and everyone needs to sleep. Can you give him something else? You don't want to bother calling the doctor, but when you try to read the back of the bottle, the list of active ingredients is indecipherable.

This scene is not uncommon, and it helps explain why cough and cold medications have been relabeled with "do not use in children under four." First of all, these medicines generally don't work for younger children. Some parents swear by certain OTC cough and cold treatments, but most would say they make little difference. These medicines might help for a couple of hours, but the symptoms come right back because the ingredients don't treat the actual infection. Second, the active ingredient list is so convoluted that it is unreadable. The generic names of the active ingredients have an average of fifteen letters (or five syllables), and a number of them are suspiciously similar sounding. Who can make sense of that?

Well, it turns out that not only do the ingredients look alike and sound alike but many act alike as well. So it is easy for parents to get confused and overdose their children. You might think you are giving your child a medicine for cough and sore throat (active ingredients dextromethorphan and acetaminophen) and a different medicine for fever (active ingredient acetaminophen), but you've actually given one of the ingredients—acetaminophen—twice. Or you may have tried a medicine for "severe congestion" (active ingredients guaifenesin and pseudoephedrine) and then an expectorant to help bring up the cough (active ingredient guaifenesin), and you've done it again.

Overdosing is not benign. In fact, among young children it can be downright dangerous. When the FDA first took steps to change labeling of OTC drugs in 2007, it was on the grounds that some of these medicines can cause hives, drowsiness, and problems with balance, and that overdoses of these medicines can cause rapid heart rates, seizures, loss of consciousness, and even death. Because many of these medicines were packaged and labeled for infants, parents thought they were safe. But every year about eight hundred infants and toddlers go to the emergency room with reactions to cold medicines (and as many as seven thousand children under eleven do the same). The final damning statistic was that decongestants were found to be responsible for fifty-four reported child deaths and antihistamines responsible for sixty-nine deaths between 1969 and 2006, mostly in children under age two. While this amounted only to an average of about three deaths per year, it was three deaths too many.

Within one year, the recommendation to limit the use of these medicines went from age two to age four. Many wanted it to go up to age six. The fight was led largely by a single health commissioner in Baltimore, Dr. Joshua Sharfstein. He argued that consumers often don't realize that the drug companies have no requirement to prove that these drugs are effective. As a result, parents think they are buying a medicine that is proven to work, when that isn't necessarily the case. What's more, when the medicines are actually studied, they turn out to be pretty ineffective. Any risk is unacceptable, but especially if there is no benefit to taking the medicine in the first place.

There is still a huge portion of the pediatric population who can take OTC cough and cold medicines safely. Technically, this includes all children over age four, though I tend to agree with Dr. Sharfstein that six is a safer minimum age. But there are millions of school-age tween and teenage kids who need to get back to school and want to use a medicine to get well faster. For this group, it helps to understand how each of the active ingredients in these various medicines works.

Decongestants take the excess water out of the nose and sinuses— essentially dehydrating the entire area. They work by constricting the blood vessels in the nose, throat, and sinuses, reducing mucus production and removing the sensation of fullness in the nose. Decongestants include pseudoephedrine, phenylpropanolamine, and phenylephrine. I find that decongestants are most helpful when the nose is running clear. If your child has a lot of thick green mucus and you give him a decongestant, the mucus often just becomes thicker as water is diverted away. In this scenario, it is not uncommon to see a child with thick mucus in the nose wind up with a sinus infection—the mucus is so thick that it plugs the sinuses, creating a new problem.

Antihistamines are also drying agents, but they work by blocking the action of histamines, the body's allergic middlemen. For this reason, they are particularly effective at treating allergic symptoms like watery eyes and sneezing. The most common antihistamines in OTC medications are diphenhydramine, chlorpheniramine, and brompheniramine. The same rule that applies to decongestants tends to apply to antihistamines: when the mucus is thick, they can dry a little too well and cause a secondary problem. Antihistamines also tend to cause significant sleepiness for many users. This creates an issue when kids are trying to use the medicine during the daytime in order to get back to school more quickly.[1]

Antitussives stop coughing. They are most often found as dextromethorphan on the active ingredient list. Something that stops a cough sounds great in theory. But in reality, we cough for a reason: to get the infection and its accompanying inflammation and mucus out. In fact, if you suppress your child's cough too well, the infection can settle, turning into bronchitis or even pneumonia. Antitussives are best when the cough is so frequent that it is causing rib pain or muscle aches, or at the tail end of an illness when an annoying cough is all that is left (but the infection that started it all is long gone).

I love expectorants. They sound completely counterintuitive because they make you cough. But think about it: for the same reason that antitussives are often bad, expectorants are great. You cough everything up and move it out. The active ingredient is usually guaifenesin. Of all the medicines listed in this chapter, guaifenesin really is the safest—it's almost impossible to overdose on it.

Fever reducers come in two forms: acetaminophen and ibuprofen. Acetaminophen (in Europe, they call it paracetamol) is synonymous with Tylenol. Ibuprofen is the active ingredient in Motrin and Advil. Acetaminophen tends to show up in many of these OTC cough and cold products; ibuprofen rarely does.

No matter how old the user, everyone needs to be cautious about mixing medicines. Learn how to read labels and take the extra minute or two to make sure that you are not giving your child (or yourself) a double dose of any one active ingredient. If you do accidentally give your child (or take for yourself) too much of any one thing, the American Association of Poison Control Centers[2] is a phenomenal resource. Representatives are available twenty-four hours a day and can help you to determine whether you need to go to the emergency room or just stop panicking.

WHAT IS THE BOTTOM LINE?

So what does a parent with a sick child do? If it's a common cold, the best treatment is really no different from what your own mother (and her mother and so on) recommended: rest, fluids like good old chicken soup, and time to let it pass. You can feel vindicated in your decision to avoid the OTC cough and cold medicines because they won't help that much anyhow.

However, if you've got teenagers who cannot bear to miss school, or if your ten-year-old is at the tail end of her illness and she desperately wants to play in that basketball game, you can try an OTC medicine. Just use caution and follow some simple rules: always read the dosing instructions on the bottle, only give one medicine at a time, and consult a health-care professional if you think you need additional medication. Also, it's important to recognize that OTC medicines only mask the symptoms—the virus has to run its course.

As for the parents of toddlers who have kids that seem to be sick every other week? Sorry. That's the way it goes in preschool. We build our immunity by getting infections and then warding them off. All the decongestant in the world isn't going to stop that.

Cough and cold medicines aren't glaringly dangerous. But they aren't 100 percent safe either. For that reason the FDA requested, and the medicine manufacturers agreed, that labels be changed on the bottles. Don't lose sleep over the fact that you gave your child a decongestant when she was a baby—there are no long-term implications. The point of changing the labels was to remove any potential future risk. But if you couple the risks with a set of medicines that don't really work very well in the first place, it makes sense to just skip them for the little kids.

WHAT'S IN MY HOUSE?

You guessed it: expectorant. I use OTC medicines very rarely, but when I do, my kids are pretty much limited to expectorants and fever reducers. A family friend with chronic sinusitis recently told me that she needs something more to relieve her congestion. I bought her a neti pot, which is available in most drugstores. With this contraption, you literally rinse your sinuses by pouring salt water up your nose. No kid will ever go for it, but for teens and adults, it works better than almost any medicine.

Chapter 25

Vaccines

WHAT IS THE QUESTION?

What *isn't* the question about vaccines? Over the past few years, vaccines have become a source of great debate among doctors, scientists, and parents. Vaccines are blamed for everything from autoimmune disease to neurological disorders, but the primary concern is their possible association with autism. Some people question the number of vaccines given at one time; others suspect the preservative used to keep them sterile; and still others worry about the young age at which babies receive these shots.

So what is the bottom line on vaccines? Are they dangerous? How about when more than one vaccine is given at a time? Or when they are given to a newborn baby? There is so much bad press—could vaccines possibly be safe?

WHAT IS THE DATA?

There is a tremendous amount of controversy surrounding vaccines (also known as immunizations, or sometimes simply "shots"). But there is an equally great—if not greater—amount of misinformation. So I will begin by reviewing the basics, including the number of recommended immunizations and the standard schedule. Only then can we dig our teeth into the issues.

There are eleven different immunizations recommended between birth and six years. Each immunization must be given one, two, three, or four times, so the actual number of doses recommended by the **AAP** and the **CDC** is thirty-six. Now, don't get too carried away by the number thirty-six, because many of these shots come in combination. So the typical child will be stuck by twenty to thirty needles over the first six years of life, depending on what combinations of vaccines are given.

Between kindergarten and college, there are booster shots and, for girls, a new vaccine in a series of three shots. If a boy received all the prekindergarten immunizations he was supposed to get according to the AAP schedule, then between seven and eighteen years he would need an additional two; for a girl, an additional five. On top of these, the influenza vaccine is recommended for all children under age eighteen, every year. So depending upon how you count it, through grade school and high school a child will receive anywhere from two to seventeen immunizations. Is your head spinning yet?

The point of this chapter is not to give a tutorial on what shots are needed when. If you want to know the official schedule, the AAP and CDC Web sites have it prominently displayed. Besides, the schedule changes constantly, and so even if I were to include it, the information would be outdated by the time this book is on the shelf.

It's also not the objective of this chapter to go through the history of each vaccine and to explain why a child should (or shouldn't) get each one. There are excellent books devoted entirely to this topic. I certainly believe that parents should educate themselves as to why each vaccine was developed—it is incredibly helpful to learn about the illness that a particular vaccine prevents. But this chapter is about something different: it is about whether vaccines are dangerous or safe.

In medical science, there is correlation and there is causation. Correlation simply pairs two ideas. For instance, there is a strong correlation between the number of cavities found among grade-school children and the size of their vocabulary. But, as we like to say in medicine, correlation is not causation. Causation means cause and effect—it means that the presence of one variable is responsible for the presence of the other. The presence of cavities among kids in grade school does not cause them to have greater vocabularies. Rather, both cavities and vocabulary size are a function of age: the older a child, the more likely he is to develop cavities and the more likely he is to master a large vocabulary. But one didn't cause the other; the two are simply correlated. The issue of correlation versus causation is a major sticking point in the debate over vaccines. One of the main reasons why parents are so confused is that vaccines can be correlated with illness—particularly autism and developmental delay—but no one has proven causation.

If vaccines are in fact correlated with certain illnesses among children, why would you ever choose to give them? How can you be sure vaccines don't *cause* autism? In other chapters in this book one of the prevailing arguments is that if something could possibly be associated with a negative outcome, then it is harmless to avoid it. Since a cell phone's EMF may be associated with brain cancer, kids should avoid exposure. Since **genetically modified food** products may have an effect on growing bodies, when possible kids shouldn't consume them. Since the body can absorb **BPA**, even though we don't know what effect it may have we should try to stay away from plastics containing it. So the same should hold true for vaccines, right? Wrong.

There are two major flaws with applying the avoid-if-you-can philosophy to vaccines. First, avoidance is not harmless. There is a consequence to avoiding immunizations: the risk of the disease the vaccine was meant to protect against. The eleven recommended vaccine types protect against fifteen different infections. The majority of these can be life threatening to infants and children. This is not a small detail; it is a massively important fact. *Before vaccines were developed, many of these infections killed children by the thousands.* Therefore, whether to vaccinate is as much a choice about preventing potentially life-threatening illnesses as anything else.

The other major problem with applying the avoid-if-you-can approach to immunizations is that their utility changes over time. Some vaccines are critically important for babies but much less so for toddlers and older children; others are unnecessary (or unavailable) for infants but lifesaving for older children. Therefore, depending on the age of a particular child and the specific type of vaccine, the answer to the question about avoidance will change.

So vaccines aren't categorically avoidable. It simply isn't possible to decide not to give them without rethinking that decision over and over again: what is the chance my child will get a vaccine-preventable disease if I don't immunize? But if I do immunize, is the vaccine dangerous? Will it cause autism? These are the $64,000 questions that come up at each and every checkup visit with your pediatrician, no matter where you stand on the spectrum.

The first—and certainly in my own practice most common—concern about vaccines is that one shot can "change" a child. Many who believe that immunizations are dangerous argue that there is a temporal association between vaccines and various diagnoses. These parents say that within hours or days or weeks of receiving a vaccine, their child changed: he lost speech, or no longer made eye contact, or began behaving differently, or was less coordinated, or in some other developmental way seemed to take a step backward.

I have never discounted this anecdotal association—parents know their child best and can usually pinpoint the moment when something shifted. Here's the problem: humans are cause-and-effect beings. We seek explanations for everything in our world. That's why you are reading this book: so you can understand what might endanger your children and avoid those things like plagues. Therefore, when faced with any diagnosis—especially one like autism—it is human nature to search for the cause. And when a parent scrolls through the potential causes—namely, things she gave to her child—an immunization in a syringe with its long needle and massive negative PR will naturally jump to the top of the list.

But if you allow yourself to think along these lines, you have to weigh *all* the other potential causes that may have affected that child: the air we breathe, the food we eat, the plasticizers everywhere in our homes (not just in baby bottles but in vinyl and cosmetics and toys), the flame-retardant chemicals coating our electric devices and mattresses, the synthetics in our clothes and carpets, the virus that caused a cough and cold, the medicines you gave your child when she was sick, and on and on. You also have to consider the things you don't see: the exposures outside your home—whether at school or on playdates or at gym class. It is not fair to blame vaccines without also pointing a finger at every other variable that may have factored into a child's development and health. This type of thinking makes the puzzle almost unsolvable.

Despite the massive scope of this mystery, there are people working hard to solve it. Scientists are furiously trying to figure out whether any of these dozens of potential variables could be contributing to the epidemic rise in autism we are seeing. But because there is so much attention on immunizations, and such a wide sense of cynicism, so far the vast majority of research has focused on vaccines themselves. Over the past decade, parents have become increasingly skeptical of vaccines and in higher and higher numbers are choosing not to give them. The idea behind focusing the research effort on vaccines is that if scientists can prove that vaccines are safe, parents will be reassured and children will be immunized against very serious infections. Then researchers can move on to other possible underlying causes of autism and developmental delay.

Most pediatricians—and certainly their governing body, the AAP—are satisfied that studies have proven the safety of each and every vaccine. The studies continue, but findings to date are very reassuring. This is not a conspiracy to convince parents to immunize but a group of physicians and researchers dedicated to the health and well-being of children. These are people who have given years of their lives to trying to understand whether vaccines cause harm and ultimately have concluded that they do not. If vaccines looked to be dangerous, this would be the first group to tell you so.

These scientists have even scrutinized the additives in vaccines, like thimerosal (the infamous mercury-containing preservative).[1] In fact, as a testament to how aggressive the AAP will be in an effort to protect children, the organization voted to remove thimerosal from all vaccines (with the exception of some forms of flu vaccine) effective in early 2001. This decision was not based on data that thimerosal caused developmental problems—rather, it was based totally on theoretical arguments. It was not an unreasonable theory, as mercury can cross the blood-brain barrier and settle among the neurons of the brain. The AAP wasn't willing to risk exposing millions of children to something that even *might* be dangerous.

Sadly, the rate of autism has continued to rise at a steady pace over the years following the mandatory removal of thimerosal. Many say that this simply proves the fact that the additive never caused autism in the first place and that this one variable can be taken off the list of possible causes. But the AAP stands by its decision to remove mercury from the vaccines because the organization understood that if there was a chance the preservative could cause problems and there was a way to eliminate that risk, it should be removed.

I think as time goes on the thimerosal argument will continue to fade. But still, as far as many medicine outsiders are concerned, vaccines remain high on the list of likely causes of autism. Just because the mercury preservative may be vindicated, some critics argue, vaccines aren't necessarily safe. In fact, for years parents and advocacy groups supporting children with autism have suggested that it's not just the preservative in vaccines but also the number of immunizations given at any one time that may be the tipping point for some autistic children. Their general argument is that vaccines stimulate the immune system, and so when multiple immunizations are given at one time, it stands to reason that the immune system will be more stimulated. Some parents (and physicians) contend that if immune stimulation triggers autism or developmental delay in a domino-like effect, then more stimulation is more likely to result in autism. The solution they propose is to stagger the shots.

A staggered schedule means that fewer vaccines are given at any one time. If a family wishes to keep the immunizations up to date—meaning that the parents plan to give the recommended vaccines by a certain age, but just don't want to give them all at once—then those parents will bring their child back to the doctor's office for vaccines between regular checkup visits. Many parents who stagger will request one vaccine at a time, visiting the pediatrician's office as often as every four to six weeks. Staggering is a reasonable solution to some parents' concerns: if a child is able to get one vaccine every month, it is possible to keep that child relatively up to date according to the AAP schedule. But staggering is not without its negatives.

The biggest downside to staggering is that a child comes to the doctor's office every few weeks for a shot. Over time, the office becomes a frightening place because almost every time that kid walks into the office he will get a shot. This is very anxiety provoking, not to mention time consuming as parents have to come back to the office frequently.

Another big negative is that a staggered schedule—especially one where a single shot is given every month—often doesn't allow for the child to get sick. A toddler gets colds and coughs an average of six to eight times per year. This is normal. If a parent is trying to stick to a strict shot schedule but a child has a cold, that month's shots will have to wait. So even when parents are very conscientious about trying to maintain a staggered shot schedule, children fall behind much of the time anyway.

The final negative is that, as it turns out, there is no data to back up the premise that a staggered schedule is less overwhelming for the body. Here's why. Staggering is based on the idea that fewer vaccines at one time can be better for the body. But when you walk down the street or eat in a restaurant, your immune system is bombarded with dozens, maybe hundreds, of organisms that it recognizes as foreign. We are designed to handle this onslaught; multiple viruses and bacteria and fungi coming our way should not cause a healthy immune system to go on overload. Likewise, multiple vaccines at once shouldn't overload the immune system either.

The counterargument is that the vaccines aren't the same as random exposure to the bacteria and viruses on the subway or in an elevator or on the schoolyard. Rather, the vaccines contain a certain number of immune stimulants loaded into a syringe and then injected into the body. This, vaccine critics might say, is the immune system equivalent to the atomic bomb.

Not true, reply the scientists (and me as well), because the vaccines aren't actual live infections. Rather, most are made from copies of one corner of the infection—purified proteins designed to "trick" the immune system into thinking the body is infected with a virus or bacteria. There is no true infection.[2]

At this point, I think it's worth taking a break and looking at how the majority of vaccines are made. Without understanding the mechanics of vaccine manufacturing, it can be impossible to decide whether or not to stagger your child's shots. Think of an infection as a letter in an envelope. Scientists who make vaccines first take the envelope and discard the letter inside. Then they tear off a small piece from the corner of that envelope; the rest is thrown away. Next the scientists make a copy of the corner of the envelope, and it is that copy (of the corner) that is used in the immunization. It isn't even the real envelope. That corner represents a protein on the outer shell of an infection. When the immune system sees the copy of the protein from the bacteria's outer shell (the corner of the envelope, if you will) it is tricked into thinking that the real infection is present. The protein, meanwhile, is a purified product. It stimulates the immune system to mount a response, but it doesn't make the body sick the way the infection would. So if and when the real infection comes along, the body is primed and ready to fight it.

Now take that rudimentary explanation and apply it to the concept of receiving multiple vaccines at once. If the products are so pure and benign, they shouldn't create massive problems—or any problems, for that matter—for the immune system, regardless of whether there is one purified protein or more than one.

The proteins in vaccines weren't always quite so purified. In fact, when we were children the original **DPT** vaccine contained whole-cell pertussis (the equivalent of the envelope with the letter inside), and it was this whole cell with its approximately three thousand different antigens that was responsible for fevers above 105°F in some recipients and seizures that accompanied those fevers. Today's **DTaP** utilizes acellular pertussis, missing the whole cell and containing only a little corner protein with no more than five antigens.

Some people bristle at the idea of giving their child more than one immunization in a syringe at any one time. For instance, they want to separate the components of the **MMR** into a measles vaccine, a mumps vaccine, and a rubella vaccine. It is important to understand that getting one vaccine at a time isn't any safer. The data confirms this. Study after study documents that receiving multiple immunizations at once—whether mixed together in a solution and given in a single injection, or given in separate shots over several minutes or several months—does not affect the immune system's reaction. There is no increase in the number or intensity of side effects, including fever, diarrhea, seizure, and other neurological symptoms. Studies also show that the response to the vaccines—the immune system's ability to form antibodies following vaccine administration—is no different if a child receives a solo vaccine in a single injection or a combination vaccine or multiple vaccines in multiple syringes.

Now, I have said that staggering itself—the act of giving a few or even just one vaccine at a time—isn't dangerous, but the upshot of staggering may be. By giving vaccines a few at a time, a child is left unprotected against some vaccine-preventable illnesses. So there is actually danger to staggering if a child is exposed to an infection before being vaccinated against it.

What about giving vaccines to newborns? Does a brand-new baby really need to be immunized within the first few hours of life? The answer is yes for some children and no for most. The children who need vaccines immediately are those whose mothers are carriers of the hepatitis B virus. These infants have been exposed to an infection that can be transmitted during delivery; receiving the vaccine just after birth (in combination with **HBIG**, the hepatitis B antibody) can spare them lifelong illness that includes profound liver disease. For children of mothers who tested negative for hepatitis B during pregnancy and who have no risk factors for acquiring it during pregnancy,[3] the hepatitis B vaccine at birth is much, much less important, and some say not necessary.

So to summarize, the vast majority of published scientific studies conclude that vaccines are safe; that giving them individually or in combination is not associated with increased side effects; that staggering vaccines is no safer than giving them in combination (and it leaves children unprotected against some diseases for a longer period of time); that their mercury preservative (now gone) was not the trigger for autism; and that the rate of autism continues to rise independent of all of this, suggesting that there is a reason other than vaccines underlying the epidemic.

But still—maybe more now than ever—there is a great divide. With scientists and (many) doctors on one side and doubting parents or parents of children with disabilities on the other, we have reached something of an impasse. One group believes strongly that vaccines do not burden the immune system and do not trigger life-altering illnesses like autism; the other group disagrees. And then there's a third group caught in the middle: parents with too little time to do all the research themselves or no direct personal connection to the issue but with enough knowledge to be worried about vaccines. This middle group is enormous and simply isn't convinced of the safety of vaccines.

It is very difficult to persuade parents that vaccines have no impact on a child's development. This is largely because immunizations are given throughout the infant and toddler years, when children are evolving at breakneck speed and developmental issues may emerge. If a parent chooses not to vaccinate at all and a child develops autism, we can say with certainty that the autism was not caused by vaccines. But what if a parent chooses to follow the recommended schedule of immunizations—

giving them at birth, two months, four months, six months, nine months, one year, fifteen months, eighteen months, and so on? Then it is very difficult to judge that child's development independent of vaccines. By definition, developmental delay means that a newborn baby who seems perfectly normal starts to do things abnormally or doesn't begin to do what he is supposed to over the first one or two years of life. It can be an investigative nightmare trying to figure out the underlying cause, especially if the parents blame the vaccines and if the first vaccines were given when a child was just a little baby.

Study after study has concluded that the vaccines are safe and will not cause a normal child to have developmental delay or autism. But for many parents this isn't enough. Many want (and need) to find a way to vaccinate their children that feels safer. Parents want the benefit of protecting their children against scary infections without the fear of injecting them with a disability. To this end, many have settled upon staggered scheduling, spreading the vaccines apart and giving one or two at a time.

Though there is no data that staggering is any safer than giving multiple vaccines at once, as long as a child can receive the immunizations he needs at the time he needs them it is not unreasonable to give them gradually. But let me be clear that by no means do I think a staggered schedule is necessary or better, and I gave my own children multiple vaccines at one time following the AAP-recommended schedule. If you do choose to stagger, you need to speak with your doctor about the order the vaccines will be given. In my practice, I am always very clear with patients that I agree with the AAP schedule, but I am also willing to come up with an alternative approach if it means that I can give the child the most important vaccines first. It makes no sense for parents to cherry-pick the vaccines without understanding the relative importance of each one.

If you want a rank-ordering system for immunizations, here is mine. But it is worth the redundancy for me to say that my children received their vaccines according to the standard schedule and I would endorse that over this list. I also believe in evidence-based medicine, and every study I have ever read looks at effects where immunizations are spaced by at least a month. So I always recommend vaccines be given four or more weeks apart. I have heard of doctors who give them weekly, but there is no data about this approach. Therefore, when offering a staggered schedule to parents, I believe the immunizations should not be given at less than four-week intervals.

For infants, the pertussis (or "P") component of the DTaP and the pneumococcal vaccines are clear priorities. If parents are skittish about vaccines and want to immunize slowly, I always encourage them to at least give both of these at the two-month visit, the first opportunity for immunizations. If parents only want to do one, I recommend the DTaP first because serious infection from pertussis is more common, but I suspect that pneumococcus will surpass it in the next few years. Both pertussis and pneumococcus can be life-threatening infections for newborns. The immunizations need to be given three times before the first birthday and once after; the DTaP is repeated again before kindergarten. If parents truly want to protect their children against these infections, the children cannot receive just one dose but need to complete the series. This means that between two months and one year, a child needs three DTaPs and three pneumococcal vaccines. It can be hard to fit in other shots if a parent is steadfast about giving one at a time.

A close third is Hib (*Haemophilus influenzae* type B). Not long ago, this infection was the leading cause of meningitis in children. Because parents immunized more readily in the 1980s and '90s, the infection largely disappeared. I fear it will return as immunization rates decline, but for now its numbers trail pertussis and pneumococcus. Depending upon which formulation of Hib is given, a child needs two or three doses before the first birthday and one after.

Polio is also offered at the two-month visit, but for parents who are not planning to travel, this is a vaccine that is often delayed. Polio vaccine is critical in the international effort to eradicate polio from the face of the earth. But, frankly, we do not have polio in this country. So if I have a parent in my practice who is only willing to give one vaccine at a time, I need to prioritize the infections that we see in the community.[4] If you do the math, it is obvious that polio vaccine is often delayed beyond the first birthday for these children.

Rotavirus vaccine is a relative newcomer. It is started at the two-month visit and must be completed by the time a child is thirty-two weeks (eight months) old. There was a version available ten years ago that was rapidly pulled off the market because it was associated with an increase in an intestinal blockage called intussusception. The new form of the vaccine has not been linked with this complication, but it has only been on the market for a couple of years, so many pediatricians have been slow to give it to their patients. Rotavirus is the most common form of stomach virus in children. Every year, it competes with influenza for the number one spot on the list of infections that cause children to be hospitalized. Rotavirus usually just causes hideous vomiting and diarrhea, and in developed countries access to intravenous fluids can be lifesaving. However, there are cases of rotavirus with serious complications (like intestinal perforation or secondary bacterial infection, which are both potentially life threatening). In underdeveloped countries where intravenous fluids are hard to come by, the dehydration associated with rotavirus can be deadly.

Hepatitis B was covered earlier in the chapter. For children born to mothers with hepatitis B, the vaccine is critical. But if the mother doesn't have the disease and doesn't have any risk factors for the disease, this vaccine, too, can wait. I did take care of a few children whose parents chose to postpone the hepatitis B vaccine because they couldn't understand the relevance of a vaccine that protects against a disease transmitted through sex and needles. I warned that blood products carry the risk of hepatitis B, but the notion of a blood transfusion or blood product was inconceivable to them. When their children had the misfortune of acquiring illnesses that did require blood transfusions or blood products, these parents expressed great remorse about skipping the vaccine. Luckily, none of these children contracted hepatitis B from their treatments, but their parents had a rude awakening. As a result, every one of those children was suddenly visiting the office regularly to catch up on missed immunizations.

The chicken pox vaccine and the MMR are only available after the first birthday, and each is recommended a second time before kindergarten. The MMR is widely available split up into its component parts (measles, mumps, and rubella). MMR in particular has earned a hideous reputation: most parents link it with autism whether or not they have done any background research on vaccines. Some physicians indulge the parents of their patients and have them wait until the child has demonstrated normal speech or other developmental milestones before giving the vaccine. The problem here is that both measles and mumps have reappeared in this country, and these diseases are not without serious consequences. There are enough unvaccinated children today—especially in pockets of communities where vaccine skepticism is the norm—that once these viruses appear, they can and will infect hundreds if not thousands. Unvaccinated children are at risk for serious illness, including brain infection and even death. My advice is to give the vaccines. But if you must wait, certainly call your doctor and immunize the instant you hear about a case in your community.

Influenza vaccine is available anytime after the age of six months and is now recommended annually until eighteen years. Flu doesn't seem so bad until you have had it, with its 105°F fevers, shaking chills, burning muscle aches, nausea, and vomiting. Influenza is the leading cause of hospitalization for infants and the elderly, and it is responsible for thirty-six thousand deaths in this country every year. While this has long been considered an optional vaccine, I have always given it to my entire family and myself. Certainly, if your child has a compromised immune system, chronic heart disease, a genetic syndrome, or asthma, the vaccine should be given, because for these children flu can be life threatening.

The hepatitis A vaccine is probably the most common one to be delayed. It is given in a series of two doses separated by six to twelve months. The vaccine protects against a virus that can cause profound nausea and diarrhea and ultimately dehydration. In general, though, infants and children don't get very sick with hepatitis A; teenagers and adults do. So this is one vaccine that can be given at a young age—as young as a year—but really protects the older three-quarters of society.

The new **HPV** vaccine for teenage girls has received its fair share of bad press. It stings (a lot) and has been associated with a high likelihood of fainting. There are questions as to whether it works as well as advertised and how long it lasts. The issues about the HPV vaccine are really quite different from the issues addressed in this chapter, because the shot is given to teenage girls who are far beyond the window of developing autism. However, I add it for the sake of being complete, and even though the initial data appeared quite safe, there is no long-term safety data.

WHAT IS THE BOTTOM LINE?

The bottom line on vaccines is that all the data collected so far indicates they are safe.

This is a particularly hideous debate because every single person involved wants to do what is right for the child. These days, conversations about vaccines are never routine; they are emotionally intense. Parents are terrified.

When you choose whether or not to immunize your child, please consider both sides to the decision. One is the fear of a complication: will the vaccine change my child? The other—often forgotten—side is fear of the disease: if I don't vaccinate my child, what is the likelihood that she will get ill with a preventable infection? Could she die?

There is also a social-consciousness piece to this whole vaccine puzzle. We can only have the option *not* to immunize if the vast majority of people *do* immunize their children. Parents who choose not to give the MMR rely on many other parents giving it. Once we cross a critical threshold where not enough children are protected against various illnesses, the illnesses will return. And when unvaccinated children contract measles, the upshot will be awful. We have already seen pockets of spread. In 2008 alone, measles appeared in multiple states, including Arizona, Arkansas, California, Georgia, Hawaii, Illinois, Louisiana, Michigan, Missouri, New Mexico, New York, Pennsylvania, Virginia, Washington, Washington DC, and Wisconsin. If few enough people vaccinate, the community becomes unprotected against the disease. It's like a spark in dry timber: at some point the illness will spread like wildfire.

Ultimately, the decision about whether to vaccinate your child, and on what schedule, is highly personal. But remember that it is a choice in either direction: while some worry that vaccination can be harmful, I believe that not vaccinating is even more so.

WHAT'S IN MY HOUSE?

I have two children and each received all the vaccines on schedule, often several at one time. But I winced with every checkup visit, and I held my breath for the few days afterward because I had heard stories from parents in my own practice—stories of certainty that the vaccines changed their children—and the mother in me couldn't ignore them. Despite this anxiety, I went with science and chose to protect my own children against diseases that I had seen firsthand, diseases that disabled (and a few that even killed) other children.

Web Sites Used
in Researching This Book

aafp.org
aap.org
aappublications.org
about.com
ada.org
ahcpub.com
ajcn.org
ama-assn.org
americanchemistry.com
annualreviews.org
answers.com
autism.com
autisminfo.com
autismresearchcenter.com
bbc.co.uk
bevnet.com
bioinitiative.org
bisphenol-a.org
blackwell-synergy.com
blogs.wsj.com/health
blogspot.com
bmj.com
boston.com
ca.gov
cancer.gov
cancer.org
carilionclinic.org
carthage.edu
cbsnews.com

cdc.gov
cerhr.niehs.nih.gov
checnet.org
childrenshospital.org
chop.edu
cmaj.ca
cnn.com
commonground.ca
consumeraffairs.com
consumerreports.org
consumersunion.org
cornell.edu
cpsc.gov
cspinet.org
dep.state.fl.us
discovery.com
eatingwell.com
ec.gc.ca
ehponline.org
emedicine.com
energyfields.org
epa.gov
eurjcancerprev.com
ewg.org
factsonplastic.com
fda.gov
findarticles.com
food.gov.uk
foodsafety.gov

foxnews.com
fruitjuicefacts.org
ghsa.org
google.com
govlink.org
harvard.edu
howstuffworks.com
hpa.org.uk
iarc.fr
ict-science-to-society.org
idph.state.il.us
ific.org
immunizationinfo.org
jhsph.edu
jrank.org
junkscience.com
latimes.com
live.com
livescience.com
manchester.ac.uk
marksdailyapple.com
mayoclinic.com
mdconsult.com
medscape.com
milkmyths.org.uk
mnceh.org
modernmedicine.com
msn.com
myspace.com
nature.com
ncahf.org
nejm.org
nih.gov
nofany.org
npr.org
nps.gov
nrdc.org
nsc.org
nutrition.org
ny.gov
nytimes.com
oregon.gov
oregonstate.edu
organicconsumers.org
oxfordjournals.org

pediatricannalsonline.com
pediatricnews.com
phthalates.org
postitscience.com
powerwatch.org.uk
preventcancer.com
rense.com
reusablebags.com
riskometer.org
sagepub.com
sciam.com
sciencedaily.com
sciencedirect.org
senate.gov
sfgate.com
sfms.org
signsonsandiego.com
snopes.com
spacetoday.org
stanford.edu
starbucks.com
textbookofbacteriology.net
theendocrinologist.org
thegreenguide.org
thelancet.com
thinkquest.org
umd.edu
umich.edu
umm.edu
unc.edu
unisci.com
uofmchildrenshospital.org
upmc.edu
usatoday.com
usda.gov
vaccinesafety.edu
vcu.edu
washingtonpost.com
who.int
wired.com
worldwidewords.org
worsleyschool.net
wsj.com
youtube.com
ynhh.org

Top Ten *Worry Proof* Web Resources for Parents

American Academy of Pediatrics (AAP) parents' site:
http://aap.org/parents.html

CDC immunization schedule:
http://www.cdc.gov/vaccines/recs/schedules/child-schedule.htm

Cell phone driving laws:
http://www.ghsa.org/html/stateinfo/laws/cellphone_laws.html

Environmental Protection Agency (EPA) kids' site:
http://www.epa.gov/kids/

FDA recalls and safety alerts:
http://www.fda.gov/opacom/7alerts.HTML

Food, Allergy & Anaphylaxis Network Web site:
http://www.foodallergy.org/

General pediatric medical information (Children's Hospital of Philadelphia Web site):
http://www.chop.edu/consumer/your_child/index.jsp

Household products database:
http://householdproducts.nlm.nih.gov/

Poison control:
http://www.aapcc.org/DNN/

U.S. Consumer Products Safety Commission recall lists:
http://www.cpsc.gov/

Notes

INTRODUCTION

1. The statistics on guns in the home are staggering. There are 280 million guns in America. It is estimated that there is a loaded gun in one out of every ten households with children and that 12 percent of them are unlocked and just "hidden away."

CHAPTER 1. ALLERGENS

1. Tree nuts include almonds, beechnuts, Brazil nuts, butternuts, cashews, chestnuts, coconuts, filberts, ginkgo nuts, hazelnuts, hickory nuts, lychee nuts, macadamia nuts, nangai nuts, pecans, pine nuts, pistachios, shea nuts, and walnuts.
2. Most people assume that with anaphylaxis, there will be a hivelike rash. This rash only occurs about 82 percent of the time in children. As a result, children who don't develop a rash may not receive epinephrine as quickly as those who have obvious hives. Epinephrine is the most important medication to be given during a severe allergic reaction, and delayed epinephrine dosing has been associated with an increased risk of death. Therefore, it is very important for people to recognize that a child who is wheezing or vomiting after ingesting a certain food may be having a serious allergic reaction, even if there is no rash.
3. In children, the following foods account for 90 percent of food allergies: milk, egg, peanuts, soy, and wheat. In adolescents and adults, peanuts, tree nuts, fish, and shellfish cause 85 percent of food allergies.
4. One reason why anaphylaxis occurs more often outside the home is that people are more likely to consume foods made with allergenic ingredients when they are not cooking for themselves. But another, less obvious reason for this phenomenon is that epinephrine is more likely to be unavailable—at home, allergic people almost always have at least one epinephrine syringe

around, but a person on the go might have forgotten to put the medicine in a purse, backpack, or car.

5. There is some confusion as to the "official" AAP recommendation as to when to start solid foods. The AAP recommends exclusive breastfeeding until six months of age, and most of their printed information geared toward doctors emphasizes six months as a starting age for solids. But on their parent-geared Web site (www.healthychildren.org), they write, "Most babies are ready to eat solid foods at 4 to 6 months of age. Before this age instead of swallowing the food, they push their tongues against the spoon or the food. This tongue-pushing reflex is necessary when they are breastfeeding or drinking from a bottle. Most babies stop doing this at about 4 months of age. Energy needs of babies begin to increase around this age as well, making this a good time to introduce solids."

6. The one exception to this is cow's milk. The AAP is quite clear that offering cow's milk in lieu of breast milk or formula before the first birthday may result in an allergic reaction. On the other hand, many dairy products like yogurt and cheese are encouraged at around eight to ten months. This often confuses parents, but it is actually rational: dairy products (like yogurt and cheese) are better tolerated because they are fermented. This is explained in greater detail in chapter 10.

7. The insertion of antibiotic-resistance genes also has a very real effect: they resist antibiotics. It is unclear whether the antibiotic-resistance genes from the foods we eat integrate themselves into our bodies and function independently once the food is gone. If they do, we could face a significant increase in the already all-too-common problem of antibiotic resistance.

8. Despite the insistence on this fact across the Internet (on very legitimate and scientific sites), I would still urge you to speak with your allergist if you have a highly allergic child. I wouldn't encourage my peanut-allergic patients to use peanut oil.

CHAPTER 2. ARTIFICIAL SWEETENERS

1. For this same reason, it is also considered a good sugar alternative for diabetics.

2. Michael Jacobson, executive director of the Center for Science in the Public Interest, from an interview with CBS News (http://www.cbsnews.com/stories/2005/07/28/health/webmd/main712605.shtml).

3. The only documented health concern about aspartame is that people with phenylketonuira (PKU) cannot ingest it. A person with PKU lacks the ability to metabolize phenylalanine and therefore must avoid all foods and beverages with the chemical. Aspartame has phenylalanine.

4. A twelve-ounce can of diet soda has between 180 and 225 mg of aspartame; however, the majority has the lesser amount, and so most calculations that equate daily aspartame intake with soda consumption use the 180 mg/soda number.

5. Of note: between 1991 and 1995 stevia was actually considered an "unsafe food additive." It was upgraded to "dietary supplement" in 1995.

6. Technical definitions of "food additive" and "dietary supplement" can be found in the glossary.

7. From the FDA Web site's section on dietary supplements: "FDA regulates dietary supplements under a different set of regulations than those covering 'conventional' foods and drug products (prescription and Over-the-Counter). Under the Dietary Supplement Health and Education Act of 1994 (DSHEA), the dietary supplement manufacturer is responsible for ensuring that a dietary supplement is safe before it is marketed. FDA is responsible for taking action against any unsafe dietary supplement product after it reaches the market. Generally, manufacturers do not need to register their products with FDA nor get FDA approval before producing or selling dietary supplements. Manufacturers must make sure that product label information is truthful and not misleading."

CHAPTER 3. BABY FOODS

1. This refers to the Code of Federal Regulations, Title 7 (Agriculture), chapter 1 (Agricultural Marketing Service), part 205 (National Organic Program), subpart G. Translation: this is the legal definition of nonagricultural and nonorganic substances allowed as ingredients in or on processed products labeled as "organic" or "made with organic foods." What follows is a list of dozens of ingredients: http://edocket.access.gpo.gov/cfr_2003/7CFR205.605. htm.

2. This accusation is really specific to benzoates used with ascorbic-acid-containing beverages like fruit juices and some (usually fruit-flavored) sodas. The USDA, however, has stated that it considers sodium benzoate generally recognized as safe (GRAS) when used in foods at concentrations below 0.1 percent.

CHAPTER 4. FISH

1. The more oily the fish, the higher its omega content. Mackerel, herring, sardines, tuna, and salmon are all examples of oily fish with high omega-3 content. Cod, scallops, lobster, haddock, and pike have much lower omega-3 levels.

2. Many different numbers are thrown around when the topic of environmental mercury comes up. According to the FDA, every year anywhere from 2,700 to 6,000 tons of mercury are released into the atmosphere naturally (via degassing from the earth's crust and oceans) compared to an estimated 2,000 to 3,000 tons that are released by human activities, primarily burning waste (household, industrial, and especially fossil fuels such as coal).

3. Acrodynia is the disease caused by mercury poisoning in infants and children. It causes pain and redness in the hands and feet. Acrodynia has gone by many other names, including pink disease, erythredema, Swift's disease, and Feer's disease.

4. If you think about it, fetuses don't have much opportunity to get rid of many

waste products or toxins: they don't sweat or have bowel movements. This is one reason why mercury accumulates so quickly in fetuses compared to adults. The red blood cells in a fetus also hold on to mercury more steadfastly. This is not surprising: fetal red blood cells hold on to lots of things (like oxygen) with more affinity than their adult counterparts.

5. An interesting historical note: the FDA actually set a maximum allowable mercury concentration in fish at 0.5 ppm back in 1969. This illustrates quite nicely that mercury was not a new issue at all, but rather one that was recognized and underpublicized for many years. Several years later, the FDA raised that maximum to 1 ppm.

6. Most toxicologists have come to agree that the average total mercury level in a minimally exposed person is 8 parts per billion (ppb) in blood and 2 parts per million (ppm) in hair. The lowest levels known to cause symptoms in the nervous system (like tingling or numbness of the lips, fingers, or toes) are 200 ppb in blood and 50 ppm in hair.

7. PCBs, or polychlorinated biphenyls, are chemicals that were used in a variety of roles, including plasticizers, coolants, insulators, pesticide extenders, and flame retardants. They were banned from production in the late 1970s as data emerged demonstrating their toxicity. PCBs and their close cousins PBDEs are covered in detail in chapter 13.

8. Another interesting fact: canned salmon is almost always wild because farmed salmon doesn't can well.

CHAPTER 5. PROBIOTICS

1. This law, called the Dietary Supplement Health and Education Act of 1994, amended the 1958 Food Additive Amendments to the Federal Food, Drug, and Cosmetic Act (FD&C Act). The following statement is posted on the FDA Web site: " . . . As a result of these provisions, dietary ingredients used in dietary supplements are no longer subject to the premarket safety evaluations required of other new food ingredients or for new uses of old food ingredients. They must, however, meet the requirements of other safety provisions. Signed by President Clinton on October 25, 1994, the DSHEA acknowledges that millions of consumers believe dietary supplements may help to augment daily diets and provide health benefits. . . . In the findings associated with the DSHEA, Congress stated that there may be a positive relationship between sound dietary practice and good health, and that, although further scientific research is needed, there may be a connection between dietary supplement use, reduced health-care expenses, and disease prevention."

2. There is data suggesting that the same holds true for some adults who are immunocompromised or have undergone recent surgery.

CHAPTER 6. SOY

1. Phytoestrogens can be divided into four groups: the isoflavones, the coumestans, the prenyl flavonoids, and the lignans.

2. Medical literature reports that some boys—and I had one such patient in my practice—who drink soy milk and eat soy products then develop gynecomastia (inappropriate breast enlargement). The case in my practice, like those I have read about, resolved completely when the boy removed soy from his diet. So there are exceptions to this rule, but they seem to be fully reversible.

3. Here is a brief history of formula: In the United States, the first formula was produced in the early nineteenth century, with cow's milk being the basis. The rubber nipple was invented in 1845, and the first commercial formula, developed by Justus von Liebig in 1867, was called Liebig's Soluble Food for Babies. Soy formula was first manufactured and sold on a large scale in 1909. It wasn't until 1929 that soy was proposed as an alternative to cow's milk formula for infants with intolerance or "allergy."

4. Doris Rapp's *Our Toxic World* provides a much more detailed description of the step-by-step process of genetic modification.

5. Since initial publication of this book, the issue of GMOs in the EU has snowballed. Several countries within the union have been granted permission to grow GMOs by their state governments. This has generated a political firestorm, as the restrictions in Europe had previously been even more stringent about growing GMOs on home soil than they had been about importing the crops. Independent of rules about growing GMOs, some EU countries have lifted bans on importing specific crops (for instance, Austria allowed the import of GM corn in 2008).

CHAPTER 7. VITAMINS AND SUPPLEMENTS

1. The other reason is public health: if pasta is iron fortified or orange juice has vitamin D and calcium, the likelihood that people will be deficient in these nutrients goes down. Deficiencies ultimately translate into medical issues—anemia in the case of low iron; osteoporosis and depression in the case of calcium and vitamin D.

2. This happens in nature too, for instance with spinach. Here the oxalate blocks the absorption of calcium, so this particular leafy green isn't necessarily as good a calcium source as the numbers suggest. But when you are giving your child a vitamin, one would think that the manufacturer wouldn't put in two competing ingredients. When it happens naturally, it seems more forgivable.

3. Because vitamin D–fortified milk is the primary dietary source of vitamin D, some nutritionists believe that any child—or even adult—who is not consuming at least a pint of vitamin D–fortified milk per day (that's the equivalent of two eight-ounce glasses or 480 ml) should probably take a multivitamin (Ensminger 1994).

4. Since the AAP's 2008 policy statement, vitamin D drops are a bit easier to find on store shelves and online. That said, multivitamin formulations are still more widely available.

5. A legume—in case you are wondering—is technically a bean or pea plant

with the distinguishing feature that its pod has two halves. Legumes include beans, peas, and lentils, all of which are high in protein and fiber. I have taken care of my fair share of self-proclaimed vegetarians who go on a meat strike but don't really like legumes. (And some don't even really like vegetables!) These kids describe empathy for the animals that are slain, so they are making a moral dietary decision. Problem is: they can develop significant nutritional deficiencies.

CHAPTER 8. CAFFEINE

1. Though it is very difficult to drink or eat enough caffeine to have life-threatening symptoms, it is possible to overdose on caffeine pills. The potentially lethal dose of caffeine is about 10 grams. There have been reports of serious symptoms requiring hospitalization after ingestion of 2 grams of caffeine.
2. Our bone density should steadily increase from infancy through the teen years and into our twenties. Around age thirty, the process stops and actually reverses itself—from that point on, most people slowly but surely lose bone density for the rest of their lives.
3. It turns out that the coffee you choose really does have a wide range of caffeine content. Dark roasts tend to have less caffeine than light roasts because the roasting process reduces caffeine content. Percolated coffee has less caffeine (on average) than drip, and instant has the least. Robusta coffees have more caffeine than do arabica varieties.
4. The stimulation provided by energy drinks is a function of their caffeine content as well as their sugar load and other additives (like taurine). These drinks—and Red Bull in particular—are covered in more detail in chapter 11.

CHAPTER 9. JUICE

1. One hundred percent juice is more caloric than the actual fruit from which it comes simply because of how much fruit is required to make the juice: it can take up to ten oranges to make one cup of fresh-squeezed orange juice.

CHAPTER 10. MILK

1. For the sake of simplicity, in this chapter I will use the word "milk" to refer to cow's milk. All other forms of milk will be identified by their animal or plant of origin.
2. "Recombinant" means "made in a lab." "Bovine" is another word for "cow" or "cattle." GH is shorthand for "growth hormone." Basically, rBGH is artificial growth hormone for dairy cows. This acronym is used interchangeably with rBST, short for "recombinant bovine somatotropin." Somatotropin is growth hormone.
3. The topic of antibiotic resistance—including what causes it and what it means—is covered in detail in chapter 23.

4. It bears noting here that if your local coffee place is a Starbucks, it made a commitment to offer only milk from rBGH-free cows starting in early 2007.

5. Nonorganic soy and GMOs are discussed in chapter 6.

CHAPTER 11. SPORTS DRINKS, VITAMIN WATERS, AND ENERGY DRINKS

1. It can also be challenging to tell how much sugar is in a drink because sugar goes by many pseudonyms. Virtually everything with the word "syrup," "sugar," or ending in "-ose" is a sugar. This list includes most of the names by which sugar can appear: corn sweetener, corn syrup, corn syrup solids, dehydrated cane juice, dextrin, dextrose, fructose, fruit juice concentrate, glucose, high-fructose corn syrup, honey, invert sugar, lactose, maltodextrin, maltose, malt syrup, maple syrup, molasses, raw sugar, rice syrup, saccharose, sorghum, sorghum syrup, sucrose, syrup, treacle, turbinado (or turbinated) sugar, and xylose.

2. There are lots of other rehydration drinks available, many of which you can make at home. My favorite is rice water: water that contains some of the starch from rice and a pinch of salt. The carb and sodium go a long way toward turning around dehydration.

3. Along these lines, I cannot help but throw in an editorial comment about branding and marketing. It is not surprising to me that people flock to a product called Vitamin Water. We all want to believe that we can benefit our own health by eating the right foods and, yes, drinking the right water. But it amazes me that some people take it a step further, buying brands like Kabbalah Water or Liquid OM—apparently they want more than health, they want religion and philosophy as well.

CHAPTER 12. CELL PHONES AND ELECTROMAGNETIC RADIATION

1. This information was posted in 2004 on http://www.spacetoday.org; the FCC also lists data about RF bands on www.fcc.gov.

2. These theories don't really carry the titles you see here. I have given them my own names and clustered groups of theories for the sake of simplicity.

3. The terminology gets increasingly complicated here. Remember that EMF is short for "electromagnetic field"; an EMF is created when a current is supplied to an antenna and it travels out from the antenna through space. A low-level EMF has a frequency of less than 3,000 Hz. Extremely low-frequency EMFs are also called ELFs and have a frequency range of 0–300 Hz.

CHAPTER 13. FLAME RETARDANTS

1. DecaBDE was actually first banned in the European Union in 2002 with the Restriction of Hazardous Substances Directive. But the European Commission lifted the ban on its use in plastics in 2005.

2. At the end of 2009, the EPA announced that two principal US-based decaBDE

manufacturers and the biggest international importer of deca would begin to reduce the manufacture, import, and sale of the chemical in 2010 and would end sales altogether by the end of 2013. According to the EPA Web site, "EPA is concerned that certain PBDE congeners are persistent, bioaccumulative, and toxic to both humans and the environment. The critical endpoint of concern for human health is neurobehavioral effects. Various PBDEs have also been studied for ecotoxicity in mammals, birds, fish, and invertebrates. In some cases, current levels of exposure for wildlife may be at or near adverse effect levels."

CHAPTER 14. LEAD

1. In the year since the original publication of this book, the CPSC has issued another dozen lead recalls, affecting nearly 375,000 toys, dress-up jewels and purses, action figures, and balls. Almost all toy recalls on the CPSC list are for either lead exposure or choking risk.
2. The abbreviation *ug* stands for microgram; the same unit is also often abbreviated as mcg. One microgram is equal to one-millionth of a gram.
3. The Consumer Product Safety Improvement Act (CPSIA) of 2008 set new federal standards for lead and phthalates in products intended for children under the age of twelve. The maximum allowable lead concentration in surface paint was reduced to 90 ppm (by August 2009) while the maximum allowable total lead content was capped at 100 ppm (by August 2011) when feasible. While the law was passed in 2008, the lead limits were scheduled to be phased in gradually over two to three years.

CHAPTER 15. PESTICIDES

1. Personal correspondence with EPA, February 2009.
2. The most frequently used organophosphates include malathion, parathion, methyl parathion, chlorpyrifos, diazinon, dichlorvos, phosmet, tetrachlorvinphos, and azinphos-methyl. Malathion is the chemical that has been used in an attempt to control mosquito populations carrying the West Nile virus.
3. It is significant that the EWG reports neurotoxic levels of pesticides eaten by these kids, but it doesn't go into any detail describing the symptoms of these children. It would have been helpful if someone had correlated groups of children who eat high pesticide loads with specific neurologic illnesses, but to the best of my knowledge this hasn't been done. So the most we can say is that in the late 1990s more than half a million children were eating organophosphates in doses that should have created problems for their neurologic systems; whether this really happened was not documented.
4. From the Board on Agriculture and Board on Environmental Studies and Toxicology (National Research Council) 1993 statement.

CHAPTER 16. PLASTICS

1. BPA is by far the most publicly scrutinized component of plastics. However, a not-so-distant second is phthalate. This plasticizer is found in most single-use water bottles in the form of PET (resin code 1). Phthalates are covered in detail in chapter 17.

CHAPTER 17. COSMETICS: NAIL POLISH, HAIR PRODUCTS, AND PERFUME

1. Other phthalates that you might hear about include DEHP (di-2-ethylhexyl phthalate) and BzBP (benzylbutyl phthalate), both of which are used primarily in polyvinyl chloride-based plastics. These are found in household items like tablecloths, furniture, vinyl flooring, shower curtains, garden hoses, plastic clothing (raincoats), and children's toys; they are generally not found in cosmetics.
2. Testosterone is a type of androgen.
3. Congress has imposed some new limits on phthalate exposures for children. The Consumer Products Safety Improvement Act (CPSIP) of 2008, section 108, requires that three phthalates (DBP, DEHP, and BBP) be limited to a concentration of no more than 0.1 percent in children's toys intended for use by children under twelve or child-care articles for children under age three. Other phthalates (DINP, DIDP, and DnOP) have been prohibited pending further study. These new rules apply only to products intended for children (http://www.cpsc.gov/about/cpsia/faq/108faq.html).
4. The controversy over antiperspirants is covered in detail in chapter 18. Aluminum—rather than parabens—has been much more at the center of this drama.

CHAPTER 18. DEODORANT AND ANTIPERSPIRANT

1. The aluminum compounds used to make antiperspirants work include aluminum chloride, aluminum chlorohydrate, and aluminum-zirconium compounds like aluminum zirconium tetrachlorohydrex gly and aluminum zirconium trichlorohydrex gly.
2. Parabens and their related health controversy are covered in great detail in chapter 17.

CHAPTER 19. DIAPERS

1. Some cloth diaper proponents have turned this on its head and suggested that because children feel slightly less comfortable in cloth diapers, they are more likely to potty train sooner; there is not a lot of data to support this claim.

CHAPTER 20. INSECT REPELLENTS

1. Deet's actual scientific name is N,N-Diethyl-meta-toluamide.
2. This statistic comes from an article written by Gideon Koren in 2003. There

is, however, a 2004 report published by the Agency for Toxic Substances and Disease Registry on the CDC Web site reporting slightly different statistics. This report states that in children under the age of eight there have actually been thirteen cases of encephalopathy (disease of the brain), some with seizure (presumably ten) and some without (presumably three). There was one additional case of encephalopathy in an adult. The data is a bit difficult to track down, but this is simply a function of how remote the problem is.

3. According to the EPA, when applying deet to the face, first spray it on the hands and then rub it in; do not spray directly on the face.

CHAPTER 21. SUNSCREEN

1. Nonmelanoma skin cancer (which includes basal cell and squamous cell carcinoma) is the most common malignant neoplasm of the U.S. adult population, with approximately one million cases per year; more than 90 percent of nonmelanoma skin cancers can be attributed to exposure to UVB (AAP Committee on Environmental Health 1999).

2. http://www.emedicine.com/derm/topic510.htm.

CHAPTER 22. TOOTHPASTE

1. It is worth noting here how much fluoride is really in a tube of toothpaste. Most tubes have one thousand parts per million, or 0.1 percent. That equals about 1 mg of fluoride per 1 g of toothpaste. So a typical tube (150 g) has about 150 mg of fluoride. Therefore, a one-year-old would need to eat an entire average-sized tube of toothpaste to come close to the fatal dose of fluoride. Take note that toothpaste tube sizes vary widely. Prescription-strength fluoride pastes, used by people who are highly susceptible to cavities, have five to ten times the concentration of standard fluoridated toothpastes.

CHAPTER 23. ANTIBIOTICS

1. Some people would argue that the history of antibiotics dates back thousands of years. There is documentation that the ancient Chinese, Greeks, Egyptians, and Arabs all used variations of natural antibiotics (in the form of molds and plants) that were known to cure some infections. However, for the purposes of this book, I will start the clock later using the "discovery" of antibiotics to mean when the medications could be mass-produced and their mechanisms could be understood.

2. This credit is often contested because French scientist Ernest Duchesne described antibiotic effects of the *Penicillium* genus in 1897. As in many stories within the field of science, however, Duchesne's description went unnoticed and it took thirty years for someone else to describe the same phenomenon, this time with credit. Another scientist to beat Fleming to the antibiotic punch was Paul Ehrlich, a German who developed the first drug

against syphilis in 1909 (though Ehrlich's drug was a compound and was not produced by another living organism).

3. Another example of noncompliance can be seen with tuberculosis (TB) medications. The bacterium that causes TB grows so slowly that it must be treated with multiple antibiotics over several months, sometimes even years. It is exceptionally difficult to convince patients to take the medications for such a long time; even those who want to are prone to forget. As a result, 5 percent of TB strains are now completely resistant to all known treatments and have become essentially incurable.

4. For a more detailed description of antibiotic resistance genes used in genetic modification, see the allergy chapter (1) or soy chapter (6).

5. This and other economic theories on antibiotic R&D are available at the *Wall Street Journal* Health Blog (data collected January 28, 2008).

CHAPTER 24. COUGH AND COLD MEDICINES

1. In younger children, antihistamines can have the opposite effect, causing hyperactivity. This is seen in up to 25 percent of young children. Parents often come to me and ask if they can use antihistamines to help their child sleep on a long flight. I never recommend antihistamines for this purpose— they are not meant to be used as sedatives. But I also caution that if they decide to ignore my advice and do it anyhow, they better not try it for the first time when they are on the plane because there is a one-in-four chance they will have a crazy child for the entire flight!

2. The American Association of Poison Control Centers is an extremely comprehensive and accessible resource. You can look it up online (www.aapcc. org) or call twenty-four hours a day (1-800-222-1222).

CHAPTER 25. VACCINES

1. Thimerosal is a preservative that prevents bacteria from contaminating multi-dose vials of vaccine. It contains 49.6 percent ethylmercury (by weight). Note that ethylmercury is different from methylmercury. It is methylmercury that has been found to cause a variety of medical problems, ranging from birth defects to central nervous system disorders including seizures (epilepsy). Ethylmercury is *not* synonymous with methylmercury—it leaves the body much more quickly than methylmercury, and it has not been proven to cause nervous system disorders.

2. There are many types of immunizations, and these are covered in great detail in some of the references listed. But it deserves stating that there are in fact a few "live attenuated" vaccines. These are made from whole live viruses (rather than copies of parts of a virus) that have been attenuated, or weakened, so that they can no longer create an infection in the body. The chicken pox (varicella) vaccine is an example of this: you receive a tiny bit of attenuated varicella in the vaccine. Live attenuated vaccines are very effective in stimulating the immune system, but people who have compromised immune

systems—like children with AIDS or various immunodeficiencies—can actually become ill from these types of shots.

3. Mothers at risk for hepatitis B are those who have unprotected sex with more than one partner (or with one sexual partner who is hepatitis B positive), have other sexually transmitted diseases, use intravenous drugs, share a home with someone who has chronic hepatitis B, work in places with the risk of exposure to human blood products, receive hemodialysis for kidney disease, live in a correctional facility, or travel to countries with high rates of hepatitis B.

4. There has not been a case of wild-type polio in this country since 1978. This includes travelers from other countries who may have been infected with polio while they visited the United States: they may have had the infection but they did not leave it behind when they returned home. The reason is that most Americans have been immunized against polio. However, if immunization rates decline, when travelers visit the United States and bring their polio infections with them, communities with lower vaccination rates will be at increased risk of contracting the infection. Just because we haven't seen the natural form of the virus in this country for thirty years doesn't mean it cannot come back.

Glossary

AAP
American Academy of Pediatrics

ADI
Short for "acceptable daily intake," this is the amount of a food additive, residue, or chemical that can be ingested daily for an entire life span without any appreciable health risk.

ALA
Alpha-linolenic acid. This is one of the three major types of omega-3 fatty acids found naturally in foods and utilized by the body. The other two are EPA and DHA.

BPA (bisphenol A)
A chemical found in hard, clear polycarbonate plastics. BPA has been used by the plastics industry since the 1950s to make a wide variety of products, including sports equipment, medical devices, home electronics, linings in canned foods, and reusable plastic containers like baby bottles.

Certified organic
See Organic.

CFU
Short for "colony-forming unit," this is the unit used by microbiologists to standardize measurement of numbers of bacteria. A CFU is a collection of organisms presumed to have originated from one single bacterium.

CIR (Cosmetic Ingredient Review)
This cosmetics-industry-sponsored organization reviews cosmetic ingredient

safety and then publishes its results. The FDA participates in the CIR in a nonvoting capacity.

CPSC
Consumer Product Safety Commission

Deet
Short for "N,N-Diethyl-meta-toluamide," this is the most common—and among the most effective—ingredients in insect repellents.

DHA
Docosahexaenoic acid. This is one of the three major types of omega-3 fatty acids found naturally in foods and utilized by the body. The other two are EPA and ALA.

Dietary supplement as defined by the Dietary Supplement Health and Education Act (DSHEA) of 1994
A dietary supplement is (also called a nutritional supplement) "A product (other than tobacco) intended to supplement the diet that bears or contains one or more of the following dietary ingredients: a vitamin, a mineral, an herb or other botanical, an amino acid, a dietary substance for use by man to supplement the diet by increasing the total daily intake, or a concentrate, metabolite, constituent, extract, or combinations of these ingredients; is intended for ingestion in pill, capsule, tablet, or liquid form; is not represented for use as a conventional food or as the sole item of a meal or diet; is labeled as a 'dietary supplement'; includes products such as an approved new drug, certified antibiotic, or licensed biologic that was marketed as a dietary supplement or food before approval, certification, or license (unless the Secretary of Health and Human Services waives this provision)." (www.cfsan.fda.gov/~dms/dietsupp.html)

DPT
Diphtheria, pertussis, and tetanus vaccine. This version of the vaccine—also referred to as the "whole cell" version—was the form used until the mid-1990s. The DPT vaccine was associated with high fevers and seizures. A newer version of the vaccine (DTaP) became available in 1996 and since 2002 has completely replaced the DPT vaccine in the United States.

DRI (Dietary Risk Index)
A score generated by the USDA, calculated from residue levels found on a particular food and a pesticide's specific toxicity.

DTaP
Diphtheria, tetanus, and acellular pertussis vaccine. It is considered to be a safer version of the vaccine because without whole cell pertussis its side-effect profile is much more benign—high fevers and rare cases of seizures that were associated with the DPT vaccine are much less likely with this version. DTaP was first introduced in 1996; by 2002 it had completely replaced the DPT vaccine.

ED (endocrine disruptor)

The term "endocrine disruptor" represents a relatively new classification for chemicals—both natural and synthetic—that may affect the way hormones work in some people's bodies. Some EDs actually mimic hormones (for instance, a particular chemical may act the same way estrogen does within the body), while others may affect the way hormones are released or regulated.

EPA

Short for "eicosapentaenoic acid," this is one of the three major types of omega-3 fatty acids found naturally in foods and utilized by the body. The other two are ALA and DHA. (Not to be confused with *the* EPA, the American governmental Environmental Protection Agency.)

FDA

Food and Drug Administration

FD&C Act (Food, Drug, and Cosmetic Act) rules about cosmetics versus drugs

From the FDA Web site: The FD&C Act defines cosmetics by their intended use, as "articles intended to be rubbed, poured, sprinkled, or sprayed on, introduced into, or otherwise applied to the human body . . . for cleansing, beautifying, promoting attractiveness, or altering the appearance" [FD&C Act, sec. 201(i)]. Among the products included in this definition are skin moisturizers, perfumes, lipsticks, fingernail polishes, eye and facial makeup preparations, shampoos, permanent waves, hair colors, toothpastes, and deodorants, as well as any material intended for use as a component of a cosmetic product. The FD&C Act defines drugs, in part, by their intended use, as "articles intended for use in the diagnosis, cure, mitigation, treatment, or prevention of disease" and "articles (other than food) intended to affect the structure or any function of the body of man or other animals" [FD&C Act, sec. 201(g)(1)]. Deodorants are classified as cosmetics while antiperspirants are considered drugs. The FDA does not have a premarket approval system for cosmetic products or ingredients, with the important exception of color additives. Drugs, however, are subject to FDA approval. Generally, drugs must either receive premarket approval by the FDA or conform to final regulations specifying conditions whereby they are generally recognized as safe and effective, and not misbranded. Some products meet the definitions of both cosmetics and drugs (such as antiperspirant/deodorant combinations). Such products must comply with the requirements for both cosmetics and drugs.

Food additive

The following definition is from the FDA: Any substance the intended use of which results or may reasonably be expected to result, directly or indirectly, in its becoming a component or otherwise affecting the characteristics of any food (including any substance intended for use in producing, manufacturing, packing, processing, preparing, treating, transporting, or holding food; and including any source of radiation intended for any such use).

FQPA (Food Quality Protection Act)
The law passed by Congress in 1996 overhauling national pesticide standards and setting much stricter food-safety laws. Among other things, it required "an additional tenfold margin of safety" for infants and children, better health-based measures by which to gauge the safety (or danger) of pesticide residues, and proof that at a given residue level there was "a reasonable certainty of no harm" to the consumer.

Genetically modified food
A genetically modified food has had its genome (DNA) altered so that it contains one or more genes not normally found there.

Germ theory
The theory that microorganisms are the cause of many diseases. Until the mid-1800s, this idea was considered outlandish. Pioneered by a number of well-known scientists and physicians, including Louis Pasteur, the germ theory became widely accepted around the turn of the twentieth century, and this allowed for the development of antibiotics and basic hygiene practices.

GMO (genetically modified organism)
Same as genetically modified food. In a GMO, the genome (DNA) of the organism is altered using genetic engineering so that it contains one or more genes not normally found there.

HBIG
Short for "hepatitis B immunoglobulin," this injection contains antibodies against the hepatitis B virus. It is given to a newborn within twelve hours of birth when the mother is hepatitis B positive.

HPV
Human papillomavirus. The HPV vaccine is the newest immunization available to adolescents. It is designed to protect against certain types of HPV infection, which is a sexually transmitted disease known to cause cervical cancer.

IARC
International Agency for Research on Cancer

MMR
Measles, mumps, and rubella vaccine

MRSA
Methicillin-resistant *Staphylococcus aureus*

NIH
National Institutes of Health

Non-GMO

Non–genetically modified organism. Describes a food or product not on the receiving end of genetic engineering.

Organic

Organic foods are generally considered those that are grown without the use of pesticides, artificial fertilizers, human or industrial waste, irradiation, food additives, or hormones. They are non–genetically modified. Only foods carrying the label "100 percent organic" are entirely organic foods (not counting added water and salt). If the label says "organic," this means that at least 95 percent of the content by weight is organic (again, not counting added water and salt); "made with organic ingredients" means that at least 70 percent is organic. Foods that are organic or 100 percent organic may carry the "USDA Organic" seal, but foods made with organic ingredients cannot.

OTC

Over-the-counter

Parabens

These chemical preservatives help make cosmetics last longer.

ppm (parts per million)

One ppm equals one part of something per million parts of something else. This is a generic unit of measure in the sense that it can represent a lot of different units of measure (such as mg/L, ug/ml, mg/kg, lb./acre). The example given on the Water Quality Mathematical Expressions & Relationships Web site is as follows: "with respect to water quality, 50 mg Calcium/L = 50 mg Calcium per 1,000,000 mg/L of water = 50 ppm Calcium."

PBDE

Polybrominated diphenyl ether, a class of chemical flame retardant.

PCB

Polychlorinated biphenyl. These compounds—which had been used as components of products as varied as coolants, lubricants, pesticides, flame retardants, adhesives, and finishes—were banned in the 1970s due to their toxicity.

Phthalates

These chemicals, first invented in the 1930s, are plasticizers—they add flexibility and resilience to thousands of everyday commodities. Phthalates are often lumped into the category of endocrine disruptors.

Phytoestrogens

These chemicals occur naturally in plants such as soy. They look chemically similar to the hormone estrogen—hence their name. Though phytoestrogens do not behave precisely like estrogen in the human body, they are considered endocrine disruptors.

Polymer

A compound—natural or synthetic—composed of repeating structural units (monomers); also defined as a large molecule consisting of a chain of identical parts.

rBGH

Recombinant bovine growth hormone. The hormone given to many dairy cows to increase milk production.

rBST

Recombinant bovine somatotropin. Synonymous with rBGH.

Report on Carcinogens

From the Department of Health and Human Services Web site: *The Report on Carcinogens* is a "scientific and public health document first ordered by Congress in 1978 that identifies and discusses agents, substances, mixtures, or exposure circumstances that may pose a hazard to human health by virtue of their carcinogenicity [ability to cause cancer]." It is published biennially.

ug

Symbol for microgram. A microgram is one-millionth of a gram.

UV

UV is short for "ultraviolet" light. UVA and UVB are two types of UV.

WHO

World Health Organization

7CFR205.605

Refers to the Code of Federal Regulations Title 7 (Agriculture), chapter 1 (Agricultural Marketing Service), part 205 (National Organic Program), subpart G. This is the legal definition (in the form of a long list) of nonagricultural/nonorganic substances allowed as ingredients in or on processed products labeled as "organic" or "made with organic foods." (http://edocket.access.gpo.gov/cfr_2003/7CFR205.605.htm)

Works Cited

This list of works cited includes journal articles and Web references. In most chapters, I relied on multiple pages within a Web site in order to gain the maximum amount of information. During the year I spent researching and writing this book, many of the bookmarked addresses changed—a Web site would still have the information I initially identified, but the data had been reformatted or recategorized, rendering my Web addresses inaccurate. As a result, I chose to list the home page for each site rather than to give detailed addresses. My data is not hard to find because each site listed has its own search bar. I apologize, though, for the extra step you'll need to take.

INTRODUCTION: WHAT DOES "DANGEROUS" REALLY MEAN?

Nebesio, T. D., and O. H. Pescovitz. 2005. Historical perspectives: Endocrine disruptors and the timing of puberty. *Endocrinologist* 15(1):44–48.

Web sites

aap.org

blogspot.com ("Political Calculations: The Odds of Dying in the US 03/19/08")

bradycampaign.org

cdc.gov ("Surveillance for Foodborne-Disease Outbreaks—United States, 1998–2002"; "Annual Smoking-Attributable Mortality, Years of Potential Life Lost, and Productivity Losses—United States 1997–2001"; "Prevention and Control of Meningococcal Disease"; "Prevention and Control of Influenza"; "Food-Related Illness and Death in the United States"; and "Fatal Injuries Among Children by Race and Ethnicity")

nsc.org ("The Odds of Dying From . . .": http://www.nsc.org/research/odds. aspx)

nytimes.com

riskometer.org

signonsandiego.com ("Odds and Ends": http://www.signonsandiego.com/
uniontrib/20040222/news_mz1c22odds.html)

wikipedia.org

CHAPTER 1. ALLERGENS

Béliveau, S., P. Gaudreault, L. Goulet, M. N. Primeau, and D. Marcoux. 2008.
Type I hypersensitivity in an asthmatic child allergic to peanuts: Was soy lecithin to blame? *Journal of Cutaneous Medicine and Surgery* 12(1):27–30.

Bock, S. A., A. Muñoz-Furlong, and H. A. Sampson. 2007. Further fatalities
caused by anaphylactic reactions to food. *Journal of Allergy and Clinical Immunology* 119(4):1016–18.

Branum, A. M., and S. L. Lukacs. 2008. Food allergy among U.S. children: Trends
in prevalence and hospitalizations. National Center for Health Statistics Data
Brief.

Gorgievska Sukarovska, B., J. Lipozencić, and G. Stajminger. 2007. What should
we know about hypersensitivity to peanuts in topical preparations. *Acta Derma-
tovenerol Croat* 15(4):269–71.

Green, T. D., V. S. LaBelle, P. H. Steele, E. H. Kim, L. A. Lee, V. S. Mankad, L.
W. Williams, K. J. Anstrom, and A. W. Burks. 2007. Clinical characteristics of
peanut-allergic children: Recent changes. *Pediatrics.* 120(6):1304–10.

Kim, J. S. 2008. Food allergy: Diagnosis, treatment, prognosis, and prevention.
Pediatric Annals 37(8):546–51.

Koplin, J., S. C. Dharmage, L. Gurrin, N. Osborne, M. L. K. Tang, A. J. Lowe,
C. Hosking, D. Hill, and K. J. Allen. Soy consumption is not a risk factor for
peanut sensitization. 2008. *Journal of Allergy and Clinical Immunology*
121(6):1455–59.

Kumar, R. 2008. Epidemiology and risk factors for the development of food
allergy. *Pediatric Annals* 37(8):552–58.

Nordlee, J. A., S. L. Taylor, J. A. Townsend, L. A. Thomas, and R. K. Bush. 1996.
Identification of a Brazil-nut allergen in transgenic soybeans. *New England
Journal of Medicine* 334:688–92.

Rapp, D. J. 2003. *Our toxic world.* Buffalo, NY: Environmental Medical Research
Foundation.

Shah, E., and J. A. Pongracic. 2008. Food-induced anaphylaxis: Who, what, why,
and where? *Pediatric Annals* 37(8):536–41.

Sicherer, S. H., A. Munoz-Furlong, and H. A. Sampson. 2003. Prevalence of peanut and tree nut allergy in the United States determined by means of a random
digit dial telephone survey: A 5-year follow-up study. *Journal of Allergy and
Clinical Immunology* 112(6):1203–7.

Story, R. E. 2008. Manifestations of food allergy in infants and children. *Pediatric
Annals* 37(8):530–35.

Strid, J., J. Hourihane, I. Kimber, R. Callard, and S. Strobel. 2005. Epicutaneous
exposure to peanut protein prevents oral tolerance and enhances allergic sensitization. *Clinical and Experimental Allergy* 35(6):757–66.

Willers, S. M., A. H. Wijga, B. Brunekreef, M. Kerkhof, J. Gerritsen, M. O. Hoekstra, J. C. de Jongste, and H. A. Smit. 2008. Maternal food consumption during pregnancy and the longitudinal development of childhood asthma. *American Journal of Respiratory and Critical Care Medicine* 178(2):124–31.

Web sites

aap.org
cdc.gov
fda.gov
findarticles.com
nih.gov
pediatricannalsonline.com
usda.gov

CHAPTER 2. ARTIFICIAL SWEETENERS

Ludwig, D. S., K. E. Peterson, and S. L. Gortmaker. 2001. Relation between consumption of sugar-sweetened drinks and childhood obesity: A prospective, observational analysis. *Lancet* 357:505–8.

Olney, J. W., et al. 1996. Increasing brain tumor rates: Is there a link to aspartame? *Journal of Neuropathology and Experimental Neurology* 55(11): 1115–23.

Soffritti, M., F. Belpoggi, E. Tibaldi, D. D. Esposti, and M. Lauriola. 2007. Lifespan exposure to low doses of aspartame beginning during prenatal life increases cancer effects in rats. *Environmental Health Perspectives* 115:1293–97.

Web sites

cancer.gov
cancer.org
cbsnews.com
fda.gov
npr.org
umich.edu
ynhh.org

CHAPTER 3. BABY FOODS

Moore, V. K., M. E. Zabik, and M. J. Zabik. 2000. Evaluation of conventional and "organic" baby food brands for eight organochlorine and five botanical pesticides. *Food Chemistry* 71(4):443–47.

Web sites

consumerreports.org
cspinet.org
howstuffworks.com
jrank.org
sciencedirect.com
usda.gov

CHAPTER 4. FISH

Cernichiari, E., G. J. Myers, N. Ballatori, G. Zareba, J. Vyas, and T. Clarkson. 2007. The biological monitoring of prenatal exposure to methylmercury. *Neurotoxicology* 28(5):1015–22.

Davidson, P. W., G. J. Myers, B. Weiss, C. F. Shamlaye, and C. Cox. 2006. Prenatal methylmercury exposure from fish consumption and child development: A review of evidence and perspectives from the Seychelles Child Development Study. *Neurotoxicology* 27(6):1106–9.

Foran, J. A., D. H. Good, D. O. Carpenter, M. C. Hamilton, B. A. Knuth, and S. J. Schwager. 2005. Quantitative analysis of the benefits and risks of consuming farmed and wild salmon. *Journal of Nutrition* 135:2639–43.

Myers, G. J., P. W. Davidson, C. Cox, C. F. Shamlaye, D. Palumbo, E. Cernichiari, J. Sloane-Reeves, G. E. Wilding, J. Kost, L. S. Huang, and T. W. Clarkson. 2003. Prenatal methylmercury exposure from ocean fish consumption in the Seychelles Child Development Study. *Lancet* 361(9370):1686–92.

Myers, G. J., P. W. Davidson, and J. J. Strain. 2007. Nutrient and methyl mercury exposure from consuming fish. *Journal of Nutrition* 137(12):2805–8.

Oken, E., R. O. Wright, K. P. Kleinman, D. Bellinger, C. J. Amarasiriwardena, H. Hu, J. W. Rich-Edwards, and M. W. Gillman. 2005. Maternal fish consumption, hair mercury, and infant cognition in a U.S. cohort. *Environmental Health Perspective* 113(10):1376–80.

Sakamoto, M., M. Kubota, Liu XiaoJie, K. Murata, K. Nakai, and H. Satoh. 2004. Maternal and fetal mercury and n-3 polyunsaturated fatty acids as a risk and benefit of fish consumption to fetus. *Environmental Science and Technology* 38(14):3860–63.

Shannon, M. W. 2004. Methylmercury and ethylmercury: Different sources, properties and concerns. *AAP News* 25:23.

Web sites

aappublications.org
cspinet.org
eatingwell.com
ec.gc.ca
emedicine.com
findarticles.com
marksdailyapple.com
mayoclinic.com
medscape.com
nih.gov
nrdc.org
nutrition.org
senate.gov
sfms.org
umd.edu
umm.edu
worldwidewords.org

CHAPTER 5. PROBIOTICS

Gill, H., and F. Guarner. 2004. Probiotics and human health: A clinical perspective. *Postgraduate Medical Journal* 80(947):516–26.

Indrio, F., G. Riezzo, F. Raimondi, M. Bisceglia, L. Cavallo, and R. Francavilla. 2008. The effects of probiotics on feeding tolerance, bowel habits, and gastrointestinal motility in preterm newborns. *Journal of Pediatrics* 152(6):801–6.

Szajewska, H., A. Skorka, and M. Dylag. 2007. Meta-analysis: Saccharomyces boulardii for treating acute diarrhoea in children. *Alimentary Pharmacology and Therapeutics* 25(3):257–64.

Van Niel, C. W., C. Feudtner, M. M. Garrison, and D. A. Christakis. 2002. *Lactobacillus* therapy for acute infectious diarrhea in children: A meta-analysis. *Pediatrics* 109(4):678–84.

Verkler, E. 2007. Probiotics: Good bacteria? *AAP News* 28(9):1

Wanke CA. 2001. Do probiotics prevent childhood illnesses? *British Medical Journal* 322(7298):1318–19.

Web sites

aappublications.org
ific.org
medscape.com
nih.gov
npr.org

CHAPTER 6. SOY

Chen, A., and W. J. Rogan. 2004. Isoflavones in soy infant formula: A review of evidence for endocrine and other activity in infants. *Annual Review of Nutrition* 24:33–54.

Franke, A. A., B. M. Halm, and L. A. Ashburn. 2008. Isoflavones in children and adults consuming soy. *Archives of Biochemistry and Biophysics* 476(2):161–70.

Gallo, D., F. Cantelmo, M. Distefano, C. Ferlini, G. F. Zannoni, A. Riva, P. Morazzoni, E. Bombardelli, S. Mancuso, and G. Scambia. 1999. Reproductive effects of dietary soy in female wistar rats. *Food and Chemical Toxicology* 37(5):493–502.

Horn-Ross, P. L., K. J. Hoggatt, and M. M. Lee. 2002. Phytoestrogens and thyroid cancer risk: The San Francisco Bay area thyroid cancer study. *Cancer Epidemiology, Biomarkers and Prevention* 11:43–49.

Korde, L. A., A. H. Wu, T. Fears, A. M. Y. Nomura, D. W. West, L. N. Kolonel, M. C. Pike, R. N. Hoover, and R. G. Ziegler. 2009. Childhood soy intake and breast cancer risk in Asian American women. *Cancer Epidemiology, Biomarkers and Prevention*. Online: 1055-9965.EPI-08-0405v1.

Luijten M., A. R. Thomsen, J. A. van den Berg, P. W. Wester, A. Verhoef, N. J. Nagelkerke, H. Adlercreutz, H. J. van Kranen, A. H. Piersma, I. K. Sørensen, G. N. Rao, and C. F. van Kreijl. 2004. Effects of soy-derived isoflavones and a high-fat diet on spontaneous mammary tumor development in Tg.NK (MMTV/c-neu) mice. *Nutrition and Cancer* 50(1):46–54.

Nebesio, T. D., and O. H. Pescovitz. 2005. Historical perspectives: Endocrine disruptors and the timing of puberty. *Endocrinologist* 15(1):44–48.

Setchell, K. D., L. Zimmer-Nechemias, J. Cai, and J. E. Heubi. 1997. Exposure of infants to phytoestrogens from soy-based infant formulas. *Lancet* 350(9070):23–27.

Tuohy, P. G. 2003. Soy infant formula and phytoestrogens. *Journal of Paediatrics and Child Health* 39(6):401–5.

Yellayi, S., A. Naaz, M. A. Szewczykowski, T. Sato, J. A. Woods, J. Chang, M. Segre, C. D. Allred, W. G. Helferich, and P. S. Cooke. 2002. The phytoestrogen genistein induces thymic and immune changes: A human health concern? *Proceedings of the National Academy of Sciences of the United States of America* 99(11):7616–21.

Web sites

aappublications.org
annualreviews.org
blackwell-synergy.com
discovery.com
ehponline.org
food.gov.uk
nih.gov
nrdc.org
oregonstate.edu
questia.com
theendocrinologist.org

CHAPTER 7. VITAMINS AND SUPPLEMENTS

Ensminger, A. 1994. *Foods and nutrition encyclopedia.* Boca Raton, FL: CRC Press.

Willett, W. C., and M. J. Stampfer. 2001. What vitamins should I be taking, doctor? *New England Journal of Medicine* 345:1819–24.

Wooltorton, E. 2003. Too much of a good thing? Toxic effects of vitamin and mineral supplements. *Canadian Medical Association Journal* 169(1): 47–48.

Yetley, E. A. 2007. Multivitamin and multimineral dietary supplements: Definitions, characterization, bioavailability, and drug interactions. *American Journal of Clinical Nutrition* 85(1):269S–276S.

Web sites

ajcn.org
cmaj.ca
medscape.com
modernmedicine.com
nejm.org

CHAPTER 8. CAFFEINE

Barrett-Connor, E., J. C. Chang, and S. L. Edelstein. 1994. Coffee-associated osteoporosis offset by daily milk consumption: The Rancho Bernardo Study. *Journal of the American Medical Association* 271(4):280–83.

Chen, J. F., K. Xu, J. P. Petzer, R. Staal, Y. H. Xu, M. Beilstein, P. K. Sonsalla, K. Castagnoli, N. Castagnoli, and M. A. Schwarzschild. 2001. Neuroprotection by caffeine and A2A adenosine receptor inactivation in a model of Parkinson's disease. *Journal of Neuroscience* 21:RC143:1–6.

Cooper, C., E. J. Atkinson, H. W. Wahner, W. M. O'Fallon, B. L. Riggs, H. L. Judd, and L. J. Melton. 1992. Is caffeine consumption a risk factor for osteoporosis? *Journal of Bone and Mineral Research* 7(4):465–71.

De Lau, L. M., and M. M. Breteler. 2006. Epidemiology of Parkinson's disease. *Lancet Neurology* 5(6):525–35.

Dowling, J. E. 1998. *Creating mind: How the brain works*. New York: W. W. Norton & Company.

Kiel, D. P., D. T. Felson, M. T. Hannan, J. J. Anderson, and P. W. F. Wilson. 1990. Caffeine and the risk of hip fracture: The Framingham study. *American Journal of Epidemiology* 132(4):675–84.

Nathanson, J. A. 1984. Caffeine and related methylxanthines: Possible naturally occurring pesticides. *Science* 226(4671):184–87.

Packard, P. T., and R. R. Recker. 1996. Caffeine does not affect the rate of gain in spine bone in young women. *Osteoporosis International* 6(2):937–41.

Pollack, C. P., and D. Bright. 2003. Caffeine consumption and weekly sleep patterns in US seventh-, eighth-, and ninth-graders. *Pediatrics* 111(1):42–46.

Savitz, D. A., R. L. Chan, A. H. Herring, P. P. Howards, and K. E. Hartmann. 2008. Caffeine and miscarriage risk. *Epidemiology* 19(1):55–62.

Weng, X., R. Odouli, and D. K. Li. 2008. Maternal caffeine consumption during pregnancy and the risk of miscarriage: A prospective cohort study. *American Journal of Obstetrics and Gynecology* 198(3):279e1–279e8.

Web sites

aappublications.org
emedicine.com
findarticles.com
nih.gov
oxfordjournals.org
sciam.com
sciencedaily.com
usatoday.com
usda.gov

CHAPTER 9. JUICE

Kontiokari, T., J. Salo, E. Eerola, and M. Uhari. 2005. Cranberry juice and bacterial colonization in children—A placebo-controlled randomized trial. *Clinical Nutrition* 24(6):1065–72.

Nicklas, T., C. O'Neil, and R. Kleinman. 2008 Association between 100% juice consumption and nutrient intake and weight of children aged 2 to 11 years. *Archives of Pediatric and Adolescent Medicine* 162(6):557–65.

Web sites

fda.gov
fruitjuicefacts.org
idph.state.il.us
nih.gov

CHAPTER 10. MILK

Adebamowo, C. A., D. Spiegelman, C. S. Berkey, F. W. Danby, H. H. Rockett, G. A. Colditz, W. C. Willett, and H. D. Holmes. 2006. Milk consumption and acne in adolescent girls. *Dermatology Online Journal* 12(4):1.

Andiran, F., S. Dayi, and E. Mete. 2003. Cow's milk consumption in constipation and anal fissure in infants and young children. *Journal of Paediatrics and Child Health* 39(5):329–31.

Ben Halima, N., A. Krichen, M. A. Mekki, M. L. Ben, I. Chabchoub, M. Chaabouni, A. Triki, and A. Karray. 2003. Persistent forms of cow's milk allergy: Report of 6 cases. *La Tunisie Medicale* 81(9):731–37.

Chan, J. M., E. Giovannucci, S. O. Andersson, J. Yuen, H. O. Adami, and A. Wolk. 2004. Dairy products, calcium, phosphorous, vitamin D, and risk of prostate cancer. *Cancer Causes and Control* 9(6):559–66.

Chen, H., E. O'Reilly, M. L. McCullough, C. Rodriguez, M. A. Schwarzschild, E. E. Calle, M. J. Thun, and A. Ascherio. 2007. Consumption of dairy products and risk of Parkinson's disease. *American Journal of Epidemiology* 165(9):998–1006.

Host, A. 1994. Cow's milk protein allergy and intolerance in infancy: Some clinical, epidemiological and immunological aspects. *Pediatric Allergy and Immunology* 5(supplement):1–36.

Konstantynowicz, J., T. V. Nguyen, M. Kaczmarski, J. Jamiolkowski, J. Piotrowska-Jastrzebska, and E. Seeman. 2007. Fractures during growth: Potential role of a milk-free diet. *Osteoporosis International* 18(12):1601–7.

Murphy, M. M., J. S. Douglass, R. K. Johnson, and L. A. Spence. 2008. Drinking flavored or plain milk is positively associated with nutrient intake and is not associated with adverse effects on weight status in US children and adolescents. *Journal of the American Dietetic Association* 108(4):631–39.

Okada, T. 2004. Effect of cow milk consumption on longitudinal height gain in children. *American Journal of Clinical Nutrition* 80(4):1088–89.

Oliveira, M. A., and M. M. Osório. 2005. Cow's milk consumption and iron deficiency anemia in children. *Journal de Pediatria* 81(5):361–67.

Rich-Edwards, J. W., D. Ganmaa, M. N. Pollak, E. K. Nakamoto, K. Kleinman, U. Tserendolgor, W. C. Willett, and A. L. Frazier. 2007. Milk consumption and the prepubertal somatotropic axis. *Nutrition Journal* 6:28.

Web sites

ajcn.org
fda.gov
junkscience.com
mayoclinic.com
milkmyths.org.uk
nih.gov
sagepub.com
sciencedaily.com
starbucks.com

CHAPTER 11. SPORTS DRINKS, VITAMIN WATERS, AND ENERGY DRINKS

Alford, C., H. Cox, and R. Wescott. 2001. The effects of Red Bull Energy Drink on human performance and mood. *Amino Acids* 21(2):139–50.

Coombes, J. S., and K. L. Hamilton. 2000. The effectiveness of commercially available sports drinks. *Sports Medicine* 29(3):181–209.

Costill, D. L., and K. E. Sparks. 1973. Rapid fluid replacement following thermal dehydration. *Journal of Applied Physiology* 34(3):299–303.

Iyadurai, S. J. P., and S. S. Chung. 2007. New-onset seizures in adults: Possible association with consumption of popular energy drinks. *Epilepsy and Behavior* 10(3):504–8.

Maughan, R. J., L. Burke, and E. F. Coyle. 2004. *Food, nutrition and sports performance II: The international olympic committee consensus on sports nutrition*. New York: Routledge.

Nielsen, B., G. Sjøgaard, J. Ugelvig, B. Knudse, and B. Dohlman. 1986. Fluid balance in exercise dehydration and rehydration with different glucose-electrolyte drinks. *European Journal of Applied Physiology* 55(3):318–25.

Reyner, L. A., and J. A. Horne. 2002. Efficacy of a "functional energy drink" in counteracting driver sleepiness. *Physiology and Behavior* 75(3):331–35.

Scholey, A. B., and D. O. Kennedy. 2004. Cognitive and physiological effects of an "energy drink": An evaluation of the whole drink and of glucose, caffeine and herbal flavouring fractions. *Psychopharmacology* 176(3–4):320–30.

Web sites

adisonline.com
bevnet.com
sciencedirect.com
uofmchildrenshospital.org
vhi.ie

CHAPTER 12. CELL PHONES AND ELECTROMAGNETIC RADIATION

Ahlbom, A., N. Day, M. Feychting et al. 2000. A pooled analysis of magnetic fields and childhood leukemia. *British Journal of Cancer* 83:692–98.

Feizi, A. A., and M. A. Arabi. 2007. Acute childhood leukemias and exposure to magnetic fields generated by high voltage overhead power lines—A risk factor in Iran. *Asia Pacific Journal of Cancer Prevention* 8(1): 69–72.

Green, L. M., A. B. Miller, D. A. Agnew, M. L. Greenberg, J. Li, P. J. Villeneuve, and R. Tibshirani. 1999. Childhood leukemia and personal monitoring of residential exposures to electric and magnetic fields in Ontario, Canada. *Cancer Causes and Control* 10(3):233–43.

Kabuto, M., H. Nitta, S. Yamamoto, N. Yamaguchi, S. Akiba, Y. Honda, J. Hagihara, K. Isaka, T. Saito, T. Ojima, Y. Nakamura, T. Mizoue, S. Ito, A. Eboshida, S. Yamazaki, S. Sokejima, Y. Kurokawa, and O. Kubo. 2006. Childhood leukemia and magnetic fields in Japan: A case-control study of childhood leukemia and residential power-frequency magnetic fields in Japan. *International Journal of Cancer* 119(3):643–50.

Leszczynski, D., S. Joenväärä, J. Reivinen, and R. Kuokka. 2002. Non-thermal activation of the hsp27/p38MAPK stress pathway by mobile phone radiation in human endothelial cells: Molecular mechanisms for cancer- and blood-brain barrier-related effects. *Differentiation* 70(2–3):120–9.

Linet, M. S., E. E. Hatch, R. A. Kleinerman, L. L. Robison, W. T. Kaune, D. R. Friedman, R. K. Severson, C. M. Haines, C. T. Hartsock, S. Niwa, S. Wacholder, and R. E. Tarone. 1997. Residential exposure to magnetic fields and acute lymphoblastic leukemia in children. *New England Journal of Medicine* 337(1): 1–7.

Sage, C. 2007 Summary for the public. *BioInitiative report*. BioInitiative Working Group.

UK Childhood Cancer Study Investigators. 2000. Childhood cancer and residential proximity to power lines. *British Journal of Cancer* 83(11):1573–80.

Web sites

answers.com
bioinitiative.org
cancer.gov
cnn.com
commonground.ca
consumeraffairs.com
energyfields.org
fda.gov
foxnews.com
ghsa.org
howstuffworks.com
hpa.org.uk
iarc.fr
live.com
livescience.com
manchester.ac.uk
myspace.com
nih.gov
nytimes.com
powerwatch.org.uk

rense.com
spacetoday.org
upmc.edu
vcu.edu
who.int

CHAPTER 13. FLAME RETARDANTS

Eisenberg, E. F. 2002 House fire deaths. *Housing Economics* November:11–13.
Eriksson, P., E. Jakobsson, and A. Fredriksson. 2001. Brominated flame retardants: A novel class of developmental neurotoxicants in our environment? *Environmental Health Perspectives* 109(9):903–8.
Fischer, C., A. Fredriksson, and P. Eriksson. 2007. Coexposure of neonatal mice to a flame retardant PBDE 99 (2,2',4,4',5-pentabromodiphenyl ether) and methyl mercury enhances developmental neurotoxic defects. *Toxicological Sciences* 101(2):275–85.
Grandjean, P., and P. J. Landrigan. 2006. Developmental neurotoxicity of industrial chemicals. *Lancet* 368(9553):2167–78. Online: doi: 10.1016/S0140-6736(06)69665-7.
Nebesio, T. D., and O. H. Pescovitz. 2005. Historical perspectives: Endocrine disruptors and the timing of puberty. *Endocrinologist* 15(1):44–48.
Schecter, A., M. Pavuk, O. Papke, J. J. Ryan, L. Birnbaum, and R. Rosen. 2003. Polybrominated diphenyl ethers (PBDEs) in U.S. mother's milk. *Environmental Health Perspectives* 111(14):1723–29.

Web sites
cdc.gov
checnet.org
cnn.com
consumeraffairs.com
cornell.edu
dep.state.fl.us
epa.gov
mnceh.org
nih.gov
nrdc.org
oregon.gov
postitscience.com
sfgate.com

CHAPTER 14. LEAD

American Academy of Pediatrics. 2005. Screening for elevated blood lead levels. *Pediatrics* 116(4):1036–46.
Farley, D. 1998. Dangers of lead still linger. *FDA Consumer Magazine* 32(1):16–20.

Web sites
aappublications.org
ca.gov
cpsc.gov
epa.gov
fda.gov
medscape.com
nps.gov

CHAPTER 15. PESTICIDES

Cohen, M. 2007. Environmental toxins and health: The health impact of pesticides. *Australian Family Physicians.* 36(12):1002–4.

Davis, J. R., R. C. Brownson, and R. Garcia. 1992. Family pesticide use in the home, garden, orchard, and yard. *Archives of Environmental Contamination and Toxicology* 22(3):260–66.

Landrigan, P. J., L. Claudio, S. B. Markowitz, G. S. Berkowitz, B. L. Brenner, H. Romero, J. G. Wetmur, T. D. Matte, A. C. Gore, J. H. Godbold, and M. S. Wolff. 1999. Pesticides and inner-city children: Exposures, risks, and prevention. *Environmental Health Perspectives* 107(S3):431–37.

National Research Council. 1993. Pesticides in the diets of infants and children. *National Academy Press*, Washington DC.

Web sites
about.com
ahcpub.com
atsdr.cdc.gov
cmaj.ca
consumersunion.org
ehponline.org
epa.gov
ewg.org
govlink.org
nejm.org
nih.gov
nofany.org

CHAPTER 16. PLASTICS

Alonso-Magdalena, P., S. Morimoto, C. Ripoll, E. Fuentes, and A. Nadal. 2006. The estrogenic effect of bisphenol A disrupts pancreatic ß-cell function in vivo and induces insulin resistance. *Environmental Health Perspectives* 114(1):106–12.

Calafat, A. M., X. Ye, L. Y. Wong, J. A. Reidy, and L. L. Needham. 2008. Exposure of the U.S. population to bisphenol A and 4-tertiary-octylphenol: 2003–2004. *Environmental Health Perspectives* 116(1):39–44.

Chapin, R. E., J. Adams, K. Boekelheide, L. E. Gray, S. W. Hayward, P. S. J. Lees, B. S. McIntyre, K. M. Portier, T. M. Schnorr, S. G. Selevan, J. G. Vandenbergh, and S. R. Woskie. 2008. NTP-CERHR expert panel report on the reproductive and developmental toxicity of bisphenol A. *Birth Defects Research (Part B)* 83:157–395.

Kamrin, M. A. 2004. Bisphenol A: A scientific evaluation. *Medscape General Medicine* 6(3):7.

Lang, I. A., T. S. Galloway, A. Scarlett, W. E. Henley, M. Depledge, R. B. Wallace, and D. Melzer. 2008. Association of urinary bisphenol A concentration with medical disorders and laboratory abnormalities in adults. *Journal of the American Medical Association* 300(11):1303–10.

Le, H. H., E. M. Carlson, J. P. Chua, and S. M. Belcher. 2008. Bisphenol A is released from polycarbonate drinking bottles and mimics the neurotoxic actions of estrogen in developing cerebellar neurons. *Toxicology Letters* 176(2):149–56.

Maragou, N. C., A. Makri, E. N. Lampi, N. S. Thomaidis, and M. A. Koupparis. 2008. Migration of bisphenol A from polycarbonate baby bottles under real use conditions. *Food Additives and Contaminants* 25(3):373–83.

Nebesio, T. D., and O. H. Pescovitz. 2005. Historical perspectives: Endocrine disruptors and the timing of puberty. *Endocrinologist* 15(1):44–48.

Schafer, T. E., C. A. Lapp, C. M. Hanes, and J. B. Lewis. 2000. What parents should know about estrogen-like compounds in dental materials. *Pediatric Dentistry* 22(1):75–76.

Soto A. M., L. N. Vandenberg, M. V. Maffini, and C. Sonnenschein. 2008. Does breast cancer start in the womb? *Basic and Clinical Pharmacology and Toxicology* 102(2):125–33.

Washam, C. 2006. Exploring the roots of diabetes: Bisphenol A may promote insulin resistance. *Environmental Health Perspectives* 114(1):A48–A49.

Willhite, C. C., G. L. Ball, C. J. McLellan. 2008. Derivation of a bisphenol A oral reference dose (RfD) and drinking-water equivalent concentration. *Journal of Toxicology and Environmental Health* 11(2):69–146.

Web sites

ama-assn.org
americanchemistry.com
bisphenol-a.org
boston.com
cerhr.niehs.nih.gov
childrenshospital.org
factsonplastic.com
harvard.edu
jhsph.edu
nih.gov
ny.gov
nytimes.com
reusablebags.com
thegreenguide.com

CHAPTER 17. COSMETICS: NAIL POLISH, HAIR PRODUCTS, AND PERFUME

Andersen, A. 2006. Amended final report of the safety assessment of dibutyl adipate as used in cosmetics. *International Journal of Toxicology* 25(supplement 1):129–34.

Barrett, J. R. 2005. Chemical exposures: The ugly side of beauty products. *Environmental Health Perspectives* 113(1):A24.

Darbre, P. D., A. Aljarrah, W. R. Miller, N. G. Coldham, M. J. Sauer, and G. S. Pope. 2004. Concentrations of parabens in human breast tumours. *Journal of Applied Toxicology* 24(1):5–13.

Harris, C. A., P. Henttu, M. G. Parker, and J. P. Sumpter. 1997. The estrogenic activity of phthalate esters in vitro. *Environmental Health Perspectives* 105(8):802–11.

Huang, P., P. Kuo, Y. Chou, S. Lin, and C. Lee. 2009. Association between prenatal exposure to phthalates and the health of newborns. *Environment International* 35:14–20.

Jackson, E. M. 2008. Subungual penetration of dibutyl phthalate in human fingernails. *Skin Pharmacology and Physiology* 21(1):10–14.

Koo, H., and B. Lee. 2004. Estimated exposure to phthalates in cosmetics and risk assessment. *Journal of Toxicology and Environmental Health* 67(23–24):1901–14.

Nebesio, T. D., and O. H. Pescovitz. 2005. Historical perspectives: Endocrine disruptors and the timing of puberty. *Endocrinologist* 15(1):44–48.

Sathyanarayana, S. 2008. Baby care products: Possible sources of infant phthalate exposures. *Pediatrics* 121(2):e260–68.

Web sites

cancer.gov
cancer.org
ehponline.org
ewg.org
fda.gov
foodsafety.gov
nih.gov
organicconsumers.org
oxfordjournals.org
phthalates.org
preventcancer.com
sciencedirect.com

CHAPTER 18. DEODORANT AND ANTIPERSPIRANT

Anane, R., M. Bonini, J. M. Grafeille, and E. E. Creppy. 1995. Bioaccumulation of water soluble aluminum chloride in the hippocampus after transdermal uptake in mice. *Archives of Toxicology* 69(8):568–71.

Darbre, P. D. 2005. Recorded quadrant incidence of female breast cancer in Great Britain suggests a disproportionate increase in the upper outer quadrant of the breast. *Anticancer Research* 25(3c):2543–50.

Darbre, P. D., A. Aljarrah, W. R. Miller, N. G. Coldham, M. J. Sauer, and G. S. Pope. 2004. Concentrations of parabens in human breast tumours. *Journal of Applied Toxicology* 24(1):5–13.

Exley, C., L. M. Charles, L. Barr, C. Martin, A. Polwart, and P. D. Darbre. 2007. Aluminium in human breast tissue. *Journal of Inorganic Biochemistry* 101(9):1344–46.

McGrath, K. G. 2003. An earlier age of breast cancer diagnosis related to more frequent use of antiperspirants/deodorants and underarm shaving. *European Journal of Cancer Prevention* 12(6):479–85.

Mirick, D. K., S. Davis, and D. B. Thomas. 2002. Antiperspirant use and the risk of breast cancer. *Journal of the National Cancer Institute* 94(20):1578–80.

Web sites
cancer.gov
cancer.org
cornell.edu
eurjcancerprev.com
fda.gov
ict-science-to-society.org
nih.gov
nytimes.com
snopes.com
unisci.com
washingtonpost.com
webmd.com
worsleyschool.net

CHAPTER 19. DIAPERS

Partsch, C. J., M. Aukamp, and W. G. Sippell. 2000. Scrotal temperature is increased in disposable plastic lined nappies. *Archives of Diseases in Childhood* 83:364–68.

Web sites
bmj.com
howstuffworks.com
stanford.edu
wired.com

CHAPTER 20. INSECT REPELLENTS

Baker, B. P., C. M. Benbrook, E. Groth, K. L. Benbrook. 2002. Pesticide residues in conventional, IPM-grown and organic foods: Insights from three U.S. data sets. *Food Additives and Contaminants* 19(5):427–46.

Fradin, M. S., and J. F. Day. 2002. Comparative efficacy of insect repellents against mosquito bites. *New England Journal of Medicine* 347:13–18.

Koren, G., D. Matsui, and B. Bailey. 2003. DEET-based insect repellents: Safety implications for children and pregnant and lactating women. *Canadian Medical Association Journal* 169(3):209–12.

Web sites
aap.org
about.com
ahcpub.com
cdc.gov
cmaj.ca
epa.gov
nejm.org
webmd.com

CHAPTER 21. SUNSCREEN

American Academy of Pediatrics, Committee on Environmental Health. 1999. Ultraviolet light: A hazard to children. *Pediatrics* 104(2):328–33.

Berwick, M. 2007. Counterpoint: Sunscreen use is a safe and effective approach to skin cancer prevention. *Cancer Epidemiology Biomarkers and Prevention* 16(10):1923–24.

Hayden, C. G., M. S. Roberts, and H. A. E. Benson. 1997. Systemic absorption of sunscreen after topical application. *Lancet* 350(9081):863–64.

Rigel, D.S., R. J. Friedman, and A. W. Kopf. 1996. The incidence of malignant melanoma in the United States: Issues as we approach the 21st century. *Journal of the American Academy of Dermatology* 34(5):839–47.

Sollitto, R. B., K. H. Kraemer, and J. J. DiGiovanna. 1997. Normal vitamin D levels can be maintained despite rigorous photoprotection: Six years' experience with xeroderma pigmentosum. *Journal of the American Academy of Dermatology* 37(6):942–47.

Twombly, R. 2003. New carcinogen list includes estrogen, UV radiation. *Journal of the National Cancer Institute* 95(3):185–86.

Web sites
aappublications.org
carilionclinic.org
emedicine.com
mdconsult.com
nature.com
oxfordjournals.org

CHAPTER 22. TOOTHPASTE

Goldman, A. S., R. Yee, C. J. Holmgren, and H. Benzian. 2008. Global affordability of fluoride toothpaste. *Globalization and Health* 4:7.

Web sites

ada.org
cdc.gov
emedicine.com
ewg.org
msn.com
ncahf.org
nih.gov
nytimes.com

CHAPTER 23. ANTIBIOTICS

Boyles, S. 2007. More U.S. deaths from MRSA than AIDS: In 2005, more than 18,000 deaths attributed to MRSA, CDC reports. *WebMD Health News*. 2007. webmd.com.

Nelson, R. 2003. Antibiotic development pipeline runs dry. *Lancet* 362(9397):1726–27.

Todar, K. 2008. Bacterial resistance to antimicrobial agents. *Todar's Online Textbook of Bacteriology*. textbookofbacteriology.net.

Web sites

aappublications.org
bbc.co.uk
textbookofbacteriology.net
thelancet.com
thinkquest.org
unc.edu
wsj.com

CHAPTER 24. COUGH AND COLD MEDICINES

American Academy of Pediatrics. 2007. Use of codeine- and dextromethorphan-containing cough remedies in children. *Pediatrics* 99(6):918–20.

Morbidity and Mortality Weekly Report. 2007. Infant deaths associated with cough and cold medications. CDC Podcast March 9. cdc.gov.

Zwillich, T. FDA mulls limits on kid's cough medicine. 2008. webmd.com.

Web sites

aap.org
aappublications.org
cdc.gov
chop.edu
nih.gov
webmd.com

CHAPTER 25. VACCINES

Baker, J. P. 2008. Mercury, vaccines, and autism: One controversy, three histories. *American Journal of Public Health* 98(2):244–53.

Halsey, N. A., and S. L. Hyman. 2001. Measles-mumps-rubella vaccine and autistic spectrum disorder: Report from the new challenges in childhood immunizations conference convened in Oak Brook, Illinois, June 12–13, 2000. *Pediatrics* 107(5):e84.

Shannon, M. W. 2004. Methylmercury and ethylmercury: Different sources, properties and concerns. *AAP News* 25:23.

Thompson, W. W., C. Price, B. Goodson et al. 2007. Early thimerosal exposure and neuropsychological outcomes at 7 to 10 years. *New England Journal of Medicine* 357:1281–92.

Waldman, M., S. Nicholson, N. Adilov, and J. Williams. 2008. Autism prevalence and precipitation rates in California, Oregon, and Washington counties. *Archives of Pediatric and Adolescent Medicine* 162(11):1026–34.

Web sites
aafp.org
aap.org
aappublications.org
ama-assn.org
autism.com
autisminfo.com
autismresearchcentre.com
cdc.gov
cnn.com
immunizationinfo.org
medscape.com
nih.gov
pediatricnews.com
postitscience.com
vaccinesafety.edu
who.int
wsj.com

Index

Note: Page numbers followed by a *t* refer to tables or text boxes.